ROGERSON'S

BOOK OF
NUMBERS

BARNABY ROGERSON is an author and
publisher. Together with his partner Rose
Baring, he runs Eland Books, which specialises
in keeping the classics of travel literature
in print. He is the author of acclaimed
biographies of the Prophet Mohammed, and
the Prophet's heirs, a history of *The Last
Crusades* and travel guides to such places as
Morocco, Cyprus and Istanbul. He writes
frequently for *Vanity Fair*, *Cornucopia*, *Conde
Nast Traveller*, *Harpers and Queen* and the *TLS*.

First published in Great Britain in 2013 by
PROFILE BOOKS LTD
3A Exmouth House
Pine Street
London EC1R 0JH
www.profilebooks.com

1 3 5 7 9 10 8 6 4 2

Printed and bound in Great Britain by
Clays, Bungay, Suffolk

A CIP catalogue record for this book is available from the British Library.

ISBN 978 1 78125 0990
eISBN 978 184765 9835

The paper this book is printed on is certified by the © 1996 Forest
Stewardship Council A.C. (FSC). It is ancient-forest friendly. The printer
holds FSC chain of custody SGS–COC-2061

ROGERSON'S
BOOK OF
NUMBERS

BARNABY ROGERSON

P

PROFILE BOOKS

CONTENTS

Introduction vii

Millions 1
Tens of Thousands 3
Thousands 7

Hundreds 12
Hundred and... 18

Hundred to Ninety 22
Eighties 28
Seventies 31
Sixties 41
Fifties 44
Forties 50
Thirties 58
Twenties 65
Nineteen to Eleven 78

Ten 131
Nine 138
Eight 148
Seven 158
Six 186
Five 197
Four 214
Three 235
Two 256
One 267
Zero 270

Introduction

This BOOK OF NUMBERS is an array of virtues, spiritual attributes, gods, devils, sacred cities, powers, calendars, heroes, saints, icons and symbols. Its short essays try to explain the many roles numbers play in poetry, in the hierarchies of Heaven and Hell and in the many religions, cultures and belief systems of our world. And, as one reads, it becomes clear that all our supposedly separate cultures are magnificently interlinked and interrelated by a shared belief in the magical significance of numbers.

I first began to assemble the book when I was working, thirty years ago, on the decoration of an underground grotto. My menial role allowed me to watch jewellers, artists, masons and sculptors at work, creating mythical beasts from stone. The carvers, in particular, were furiously competitive. Sometimes work that had been finished the day before would be found to have been hacked off by another hand when we returned in the morning. The chief sculptor would hide when the architect came around for his weekly inspection, banned any instrument that could measure units and would not tolerate advance planning. 'Go and dream it,' he would mutter, if any of us came to him with questions. He was, however, generous with his skills and a natural teacher by example. Although I had the simplest jobs – mixing tea, making mortar and laying the floor with pebbles – he enhanced my status by asking me for an opinion on anything historical or mythological. He was especially excited by anything that had a numerical ring to it – which could help in his ideas for the decoration of the underground dome.

'Green Grow the Rushes, O' and the 'The Twelve Days of Christmas', with their twinkling, haunting symbolism, were especially well received – in part because there were a large number of different explanations to argue over.

During the lunch break we would all stretch out in the sunshine, basking in delighted contrast with our dark, underground work-zone, with its buckets of mortar, trays of sHells, mislaid tools and pools of stagnant water through which twisted sinister, serpentine lengths of power cable. As we enjoyed dreamy, picnic lunches, overlooking a beautiful castle framed by three lakes, we began to dream up a contemporary cathedral. This was not to be encrusted with minerals, sHells, distorted bricks, bleached bones, baked flints and crystals, like the structure we were building, but instead a triumphant instance of modernism. It would be a glittering shrine of glass and taut steel cables, standing alone on an island just off the ocean shore and dedicated to the Sustainer of Life. It would be approached during low tides, when visitors could walk across. The glass vaults were to be etched with thousands of images, all the names of the gods, saints and ethical teachings. At noon they would be invisible to the naked eye, but at dawn and dusk they would glow with color. The floor of this temple was to be covered with washed sand, upon which visitors could draw patterns or inscribe new versions of the wisdom of ages.

This little book is designed to aid such a work, beginning with the big numbers and working backwards.

Barnaby Rogerson, September 2013

THE
NUMBERS

Millions

MILLIONS OF ANGELS DANCING ON A PIN

The question of 'How many angels could dance on a pin' is often quoted as the essence of medieval scholasticism, a burning issue for the likes of Duns Scotus and Thomas Aquinas. In fact, although Scotus certainly troubled himself over the question of 'Can several angels be in the same place?' there is no mention of dancing on pins until it was raised as mockery in the seventeenth century by Protestant academics. Still, it's a question that ought to be answered and if we take an angel to be no more or less than an atom, then 200,000 could fit in the width of a single human hair. More impressively, neuroscientist Anders Sandberg has come up with the figure of 8.6766×10^{49} angels, based on theories of information physics and quantum gravity.

Angels dancing on pins – but how many could fit?

THE 4,320-MILLION-YEAR-LONG DAY

4,320 million of our years corresponds to but a single day in the existence of the Lord Brahma, the ultimate unifying aspect of all the teeming gods of the Hindu pantheon. For Brahma is the Absolute, the Universal Soul, Ishawara – the One Great God. This day of Brahma can be divided into a thousand units or *mah-yugas* to give us the more homely scale of 4.32 million years. They can also be multiplied up to create a Brahmic month, 259.2 billion of our years, or a Brahmic year, which is 311.04 trillion of our years, or the 100-year life-span of the supreme deity.

It is believed that we are just over halfway through the current incarnation of Brahma – in the first day of the fifty-first year of his life. It is rather like the big bang theory, but even more so, for at the end of each cycle of 311 trillion years there is an equal period of immobile darkness before the universe explodes into another Brahmic creation.

Current scientific number-crunching pitches the universe as 12 billion years of age and our earth about 5 billion years old. On the surface of this planet, plant life has been around for 500 million years, mammals have been sniffing around for 130 million years and modern man for a quarter of a million. This is in slight variance with traditional Hindu thought, which records ten primordial kings who ruled the world for 432,000 years before the Flood.

The 432 unit of measurement was always popular, not only as a reflection of the vast scales of Brahmic time but also because it was four times 108 – a number which, as we will see later, is a very propitious concept.

10s of 1000s

237,600 MILES OR 30 EARTHS

237,600 miles is the average distance between the earth and the moon, a number which suggests an intriguing inner harmony to our universe, for it is thirty diameters of the earth, sixty radii of the earth or 220 moon radii. The mystical author and numerologist John MicHell would reveal these figures with the full force of a relevation during his lectures. A self-declared 'radical traditionalist', MicHell campaigned long and hard against the destruction of England's ancient number systems in favour of the decimal system.

144,000 TO BE SAVED

This is the number who will be saved at the end of the world, as anyone who has given time to a Jehovah's Witness can attest – 12,000 from each of the twelve tribes of Israel – though the religion's more merciful interpreters declare it the number of kings who will rule over us in Heaven.

The source for this belief is Revelation 7: 4–5: 'And I heard the number of those who had received the seal. From all the tribes of Israel there were a hundred and forty-four thousand.' Verse 14 goes on to define them as the righteous, who 'alone from the whole world have been ransomed. These are men who did not defile themselves with women, for they kept themselves chaste and they follow the Lamb wherever he goes ... No lie was found in their lips; they are faultless.'

124,000 PROPHETS

The traditional number of prophets sent by God to teach the world before the coming of the Prophet Muhammad is 124,000, as recorded in a remembered saying (*Hadith*) of Muhammad, though nowadays this is considered to be of doubtful veracity. But, for those who like a definite figure, it is a useful figure to put beside the Koranic declaration, 'to every people we have sent a Prophet'. Other Islamic sources make mention of 77,000 great saints or sheikhs sent since the death of the Prophet to remind mankind of the Truth.

84,000 STUPAS OF EMPEROR ASHOKA

Mount Meru, the mythical Buddhist centre of the universe, was considered to be 84,000 Yojan units high (which makes it about 672,000 miles in elevation). This respect for 84,000 is repeated by the Jain, who measure their cycle of time in units of 84,000 years and also by belief that the Lord Buddha left behind 84,000 teachings. And so this was the number of memorial stupas that the great Buddhist Emperor of India, Ashoka, is believed to have created to hold the Lord Buddha's ashes, which he scattered across the landscape of South Asia.

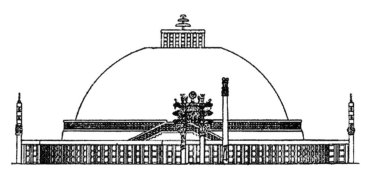

The great stupa at Sanchi, commissioned by Ashoka in the third century

10,000 BLESSINGS OF A PEACH

'Ten thousand' is poetic Chinese for 'infinite', as in 'may the Emperor reign 10,000 years' or, as it now says over the gate of Heavenly Peace (Tiananmen) in Tiananmen Square, 'May the People's Republic of China last 10,000 years.' This unit of time is symbolised by a peach, as the Chinese delight in making associations between the sounds or tonal connections of (otherwise unconnected) words. So when you look at Chinese imagery, be it an ancient watercolour or strident propaganda poster, keep an eye out for a propitious scattering of peaches, birds, bats and vases. A bird, especially a crane, has tonal connections with 'harmony', a bat with 'prosperity', a vase with 'peace' and, as we have already heard, a peach can say 10,000 years.

XENOPHON'S 10,000 MERCENARIES

Xenophon's *Anabasis* tells the story of 10,000 elite Greek mercenaries who are left isolated on the losing side of a Persian civil war and fight their way across the mountain tribes of Anatolia to reach the safety of the Black Sea coast. The history of this march in 401 BC was the original, story of swashbuckling adventure against the odds and was said to have inspired Philip of Macedon to take on the Persians. T.E. Lawrence had the book in his camel bag during the Arab revolt of 1916. And, more recently, transplanted to the gangs of New York, it became the *Warriors* video game.

10,000 New York Warriors – make their day and count 'em.

1000s

6,585 DAYS OF THE SAROS CYCLE

There are 6,585 days between one total solar eclipse and another, which is 18 years, 11 days and 8 hours. This has been known, observed and calculated for many thousands of years, but was probably first chronicled in ancient Babylon (in Mesopotamia, in modern-day Iraq). It would later be disseminated by the Greeks as the Saros cycle.

YEAR ONE – 2696 BC

The year 2696 BC used to be considered the start date for Chinese civilisation, for the winter solstice of that year was held to be the beginning of the reign of the Yellow Emperor. Most historians had accepted that the period of the Three Sovereigns and the Five Emperors is mythic time, though Huangdi, the Yellow Emperor (who comes at the end of this period), could have been based on a true historical character. Huangdi was honoured as the man who taught the Chinese how to build shelters, tame wild beasts, build boats and carts, and plant and reap the five cereals, while his wife taught weaving and silk-making, and their chief minister set out how to write, keep laws and the annual calendar.

If we were all to agree to a new world calendar system, the Chinese Year One would not be such a bad start date, for it calibrates pretty closely with other great memory pegs of world history, such as the construction of the first pyramid (2630 BC), the first era of Stonehenge (3100-2400 BC) and the first recorded king of the Sumerian city-state of Ur.

1,460 YEARS OF THE SOTHIC CYCLE OF ANCIENT EGYPT

Ancient Egypt ran a number of calendar systems, of which the simplest was a three-season farming year. The four-month seasons were Akhet (the Nile flood), Peret (growth) and Shemu (harvest) and this pragmatic division of time could be followed by anyone working in the Nile valley, though the onset of the flood could vary by up to three months.

The Egyptian priests also observed and recorded the twelve months of the lunar cycle; their twelve sets of 29-30 days fits almost but not exactly into our solar year of 365 days. The thirty days of the month were divided into three ten-day-long decan weeks, rather than our system of four seven-day-long weeks. This gave 36 decans in a year, each of which was named after a visible star. These were grouped into pairs and given a totemic spirit-animal (not unlike our zodiac) to create a succession of eighteen identities – as can be seen in one of the rings of the Dendera temple planisphere. The addition of five extra days, the *Epagomenae*, allowed the lunar cycle (12 times 30-day months) to tie in with the 365 solar year. In some cultures these odd five days were days of dread, when malignant spirits were believed to range over the earth, but the Egyptians made an annual festival of the event, and so the five great gods, Osiris, Horus, Seth, Isis and Nephthys, were all honoured in turn before New Year's Day. This customarily started on the spring equinox (21 March) so that, like the modern signs of the zodiac, the months ran from the 21st of a month to the 20th of the next.

The priests recorded important correlations between the seasons and the rising of the stars. The most famous of these was the heliacal rising or zenith of the star Sirius (known as Sopdet), which follows a very similar pattern to our solar year, dropping back just one day every four years so that every 1,460 years the zenith of Sirius's route through our skies coincides with Midsummer Day. If we believe that such stone circles at places such as Nabta Playa in Egypt can be seen of evidence of highly organised star-watching, it is possible that the first Sothic cycle was consciously witnessed around 4242 BC, followed by those in 2782 BC and 1321 BC, as well as that described in 139 AD and that observed by Kabbalists in the seventeenth century.

The zodiac depicted on the temple at Dendera.

PTOLEMY'S 1,022 STARS

The great quest of medieval science was for a perfect copy of Claudius Ptolemy's *Almagest*, written in Egypt in 147 AD. It was known to have thirteen sections, with the most accurate analysis of star and planetary paths ever achieved, alongside a catalogue of 1,022 stars, listed on a scale of magnitude of 1 to 6. It was a key that threatened to unlock the secrets of the heavens.

1,003 CONQUESTS OF DON GIOVANNI

Leporello, manservant of the fictional rake Don Giovanni (Don Juan), revealed that his master made 1,003 sexual conquests in his Spanish homeland ... as well as 640 in Italy, 231 in Germany, 100 in France and 91 in Turkey. Of course, it must be remembered that Leporello's purpose was to gently persuade Donna Elvira not to put too much trust in his master – and to amuse an operatic audience. Still, Don Giovanni's figures stack up well alongside his historic rivals. Casanova claimed to have slept with a mere 122 women. Byron (who wrote his own *Don Juan*) raced through more than 300 women (plus numerous Venetian rent boys and transvestites) before his early death in Greece, aged 36.

THE 1,001 NIGHTS

The *Kitab Alf Laylah wa-Laylah* – 'The Book of the Thousand and One Nights' – has inspired countless films, musicals and novels. The original tales are breathtakingly inventive, vulgar and discursive, full of cliff-hanger action, scented with sex, royalty and magic. Western scholars have been arguing over their origin, composition and textual tradition for some 300 years, a debate animated by the schism between

*The Forty Thieves (possibly an invention of 'The 1001 Nights' French
translator) at work. Shame they revealed their password.*

an eighteenth-century French translation of a Syrian
manuscript and a later English translation of an Egyptian
one. It seems clear that there is an ancient Persian, Indian
and Mesopotamian collection of stories at the core of 'The
Nights', which came together as a coherent whole in Arabic
in ninth-century Baghdad, was then embroidered by Iraqi
storytellers, and further embellished by tales added from the
streets, cafes and imagination of the medieval cities of Egypt,
North Africa and Syria.

Long known as 'The Thousand Nights', the collection did
not become 'A Thousand and One' until the twelfth century. Curiously, too, many of the most celebrated adventures,
such as 'Ali Baba and the Forty Thieves' and 'Aladdin and
his Lamp', were added at the very last 'textual' moment by
the first French translator (Antoine Galland), sourced from a
Maronite story-teller of Aleppo.

100s

666 – THE NUMBER OF THE BEAST

Saint John saw the beast 'rise up out of the sea, having seven heads and ten horns, and upon his horns ten crowns, and upon his heads the name of blasphemy', which seems to fit temptingly close to the old Phoenician-Canaanite myth of a sea monster Lord of Caos (Yam/Lotan) coming up out of the deep to do battle with a hero god like Baal/Hadad. In amongst the complex imagery of John's Book of Revelations, some commentators have argued that the seven-headed beast also represents the seven Roman emperors who had been responsible for the degradation of the Temple, the destruction of Jerusalem and the persecution of Judaism and its heretical offshoot – early Christianity. Counting back from John's contemporary, Domitian, these seven emperors would be Titus, Vespasian, Nero, Claudius, Caligula, Tiberius and Augustus.

But it is the 666 number that most resonates, the numerical value Saint John ascribes as the mark of the beast: 'Here is wisdom. Let him that hath understanding count the number of the beast: for it is the number of a man; and his number is six hundred threescore-and-six.' This hint at numerological

The beasts of Saint John's Revelations – as depicted by Iron Maiden.

coding allows (with different values given to each letters) that 666 would seem to identify 'Nero Caesar' when written in Hebrew (it was Nero who organised the first popular pogrom against the Christians after the great fire of Rome). 666 is also the number created when you list – or add – the first six symbols of the Roman numeral notation together, as in D (500), C (100), L (50), X (10), V (5) and I (1).

In Chinese, 666 is a tonal equivalent for 'things go smoothly' and a favoured number. It also has an alliance with the roulette table, as the sum of all the numbers on the wheel.

540 GATES OF VALHALLA

Valhalla, the Heavenly hall of feasting presided over by the Norse god Odin, was reserved for only the bravest and strongest of warriors who died a hero's death on the battlefield. It had 540 doors and was set in an immense grove of golden yew trees.

THE CINQ-CENTS

Cinq-Cents – the Council of 500 – was the elective assembly which ruled France at the end of the Revolution, between the end of the Terror and the seizure of power by General Bonaparte (1795–99). Much overlooked now, the so-called Directory period was an attempt at creating a stable and balanced democracy, with the Assembly empowered to nominate five directors, who, once they had been approved by the 250-strong Senate of 'Ancients', ruled the Republic.

The Assembly consciously looked back to the democracy of ancient Athens, which was governed through the Boule, an assembly of 500. However, the ancient model attempted to avoid the perils of influence-peddling and the factionalism of party politics by cutting out the voting process; instead, each of the ten tribes of Athens and its hinterland held a ballot to send fifty of their men to attend this standing council for a year. After a year's service, they had to resign.

THE 400

'The Four Hundred' is the nickname for the social elite of New York, an alliance of old landed families, financial speculators, manufacturers and entrepreneurs who had assimilated European social manners and snobbery in the late nineteenth century. They overlooked the divisions of the Civil War, delighted in transatlantic marriages with the nobility of Europe and guarded themselves from 'new money' coming in from the West, especially those who put too much crushed ice in their wine. The concept of the Four Hundred was popularised by Ward MacAllister, the Beau Brummell of Manhattan, who coined the expression from the number who could be comfortably entertained, and felt 'at ease', in Mrs Astor's ballroom.

365 DAYS OF HAAB

The Mayan solar calendar was divided into eighteen twenty-day months (vigesimal notation). This produced 360 days, or one tun. In common with many religious calendars of the world, the shortfall of five extra days added on to make the 365 days of the solar year was a spirit-haunted period of ill omen, the five nameless Wayab days.

The habit of multiplying by twenty continued beyond the year, so that twenty Tun (almost a solar year) is a katun, which was a great unit of time commemorated with inscribed standing stones and pairs of pyramid temples. Twenty katuns produce a baktun and twenty baktuns produce a piktun of 7,885 solar years. We have not quite got to a piktun, though quite recently, at the winter solstice of 2012, we celebrated the completion of thirteen baktuns which some observers took to be a possible date for the end of the world. The thirteen-baktun date (21.12.2012), which was safely achieved, begins by starting the calendar clock at Year One in 3113 BC.

360 DEGREES

The circle is divided into 360 degrees, which is an attempt to create a perfect universe out of our slightly wonky one. For 360 can be neatly divided by 4 to make 90-day seasons, or by 12 to make perfect months of 30 days, or by 18 to make 20-day units. This perfect ordering of the world – the exagesimal system – was codified by the Babylonians and still orders the world of geometry and time-keeping, with 60 seconds in a minute and 60 minutes in an hour.

Of course, the reality of our world was never quite as neat as those Babylonian mathematicians aspired to be, for a lunar month is actually 29 days, 12 hours and 44 minutes, not a neat 30, and a solar year (the time in which it takes the earth to orbit the sun) is actually 365 days, 5 hours and 48 minutes not a neat 360. So in the old days we made an odd thirteenth month of five days, before opting to spread them around to make some months 30 days long, some 31. And every fourth year we need our years to be 366 days long, in order to use

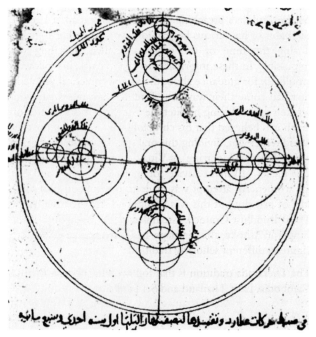

Circle rules: a fourteenth-century Arabic model for the appearances of Mercury.

up an extra day acquired by four additional units of five hours and 48 minutes.

Nonetheless, the perfection of 360 has always been aspired to, with ancient stone circles formed of 360 stones and altars formed from 360 cut stones.

VINAYA – THE 227 RULES

Vinaya are the 227 rules by which a Buddhist monk of the Theravada tradition must conduct himself – conspicuous in his orange robes, shaven head and barefoot – though he is free to disrobe himself of this obedience at any time.

As the Buddha's preaching and influence spread, it became the habit of his various beggar-followers to gather together during the time of the monsoon, when travelling was impossible. In these informal forest gatherings, his followers would ask for guidance about the various practical problems that had come their way. The Buddha's responses were not written down during his lifetime but three months after his death it is believed that his chief followers recited what they could remember, dividing this oral spiritual inheritance into either Dharma 'doctrine' or Vinaya 'discipline'. There were 227 pieces of Vinaya advice for male followers – and 311 for women. An attempt was made to order them into some sort of priority but this was abandoned. By the time of the Third Buddhist Council, assembled at the invitation of the Emperor Ashoka, this heritage had already expanded into eighteen different scholarly traditions.

The Theravada tradition is that followed in modern Burma, Cambodia, Laos, Thailand and Sri Lanka.

100 and...

114 SURA OF THE KORAN

There are 114 sura or 'chapters' within the Koran. These are named and numbered, but the names (such as The Cow or The Light) have no importance other than as a memory tag linked to some unusual feature. The chapters are not ordered by age of delivery or location (either Mecca or Medina), but in reverse length, so the shortest chapters begin the Koran and the longest end it. There is no narrative flow; indeed, at times one could almost imagine the Prophet's revelations to be addressed to 114 different types of human, for each chapter is a development or a condensation of the same essential theme: how to live and love both mankind and God.

108 STUPAS ON THE WALL

Genghis Khan's city of Karakoram, the tented capital of Asia, was encircled with a wall that was decorated with 108 stupa-shrines. This remains a highly propitious and symbolic number in Central Asia, India and the Far East. In India it is the emergency phone number, while in Japan the temples ring out the old year with a toll of 108 bell strikes, one for each of the 108 lies, 108 temptations or 108 sins resisted. The number can be satisfactorily divided into three groups of thirty-six, a third dealing with the past, a third with the present and a third with the future.

Rosaries and belts with 108 beads are also commonly worn and counted by Hindu, Zen and Buddhist monks and priests. For, linked with the list of 108 earthly moral temptations, each and every Hindu deity has 108 distinct names, titles and epithets (they seem to derive from the 54 letters of the Sanskrit alphabet, which, when recited in both their masculine and feminine forms, produces 108).

But the most beloved piece of symbolism behind the attraction of 108 seems to be in the order and shape of the numbers themselves. In Eastern philosophy, the 1 stands for the essential unity of creation; 0 for the nothingness of our future existence; and the 8 means everything; so, together, they create a chant of 'one-emptiness-infinite'.

108 NAMES OF KRISHNA

Ajanma ('Limitless and Endless') ✳ Bihari ('He Who Plays') ✳ Govinda ('He Who Pleases the Herds) ✳ Krishna ('The Irresistibly Beautiful') ✳ Murali ('Lord of Flute-playing') ✳ Navaneetha Chora ('Master of Butter-makers') ✳ Padmanabha ('Lord Who Has a Lotus for a Navel') ✳ Vardhamana ('The Formless Lord') ✳ and 100 more

Space permits listing eight rather than all 108 of Krishna's names. But 8 is also a key number for the god. Krishna is believed to be the eighth great incarnation of Vishnu: the beloved god-child, the playful flute-playing cow-herder who proves himself irresistibly attractive with his dark skin and dreadlocks. He is a model lover, muse of musicians

Krishna pleasing the herds with his flute-playing.

and actors, a great warrior hero, preacher and rescuer of women. In Western imagery we can see him as made up of identities we associate with Apollo, Pan, Dionysus, Orpheus and Christ. But behind this charming earthly identity lies the form of all-powerful Vishnu, so that many of Krishna's 108 titles will be shared by other avatars of the Supreme Being.

101 NAMES OF AHURA MAZDA

Abadah ('Without Beginning') ❈ Abee-Anjam ('Without End') ❈ Abaravand ('Detached from All') ❈ Parvandah ('Connected with All') ❈ Gel-Adar-Gar ('Who Turns Dust into Fire') ❈ Farsho-Gar ('Eternal Source of Soul Energy') ❈ and 95 more

The string of titles – again, this is necessarily a truncated selection – with which a Zoroastrian Persian once addressed Ahura Mazda is a prayer in itself. The 101 names have been preserved amongst the worldwide diaspora of the Parsee, whose safe haven has long since shifted from their original Persian homeland to the Indian state of Gujarat. A smaller list of the divine titles of Ahura Mazda is the customary starting block of the Yasna, the 72-chapter liturgy of the Zoroastrians. The name list is probably behind the old Persian practice of presenting tribute in a succession of 101 trays filled with gifts, just as a marriage gift for a bride was once expected to consist of 101 items of clothing.

101 DALMATIANS

Pongo ❈ Missis Pongo ❈ Perdita ❈ Prince ❈ and 97 pups

Dodie Smith's novel, *The Hundred and One Dalmatians*, became one of the most successful of all Disney movies – and surely it is the number of puppies that makes it so memorable. The plot begins with the kidnap by Cruella de Vil of the Dearly family's fifteen pups, which become part of a cache of 97 stolen for their fur and held captive in Hell Hall. They are ultimately rescued by the Dearly dogs, Pongo and Missis, who leave their masters in the charge of a wet-nurse, Perdita. That makes 100 dalmatians when all the pups make it back. Number 101 turns out to be Perdita's long-lost love, Prince.

100

A hundred is a ubiquitous element of power and finance. If ancient Greek gods were angered, they could be appeased with the blood bath of a hecatomb – the sacrifice of 100 oxen. A hundred was also long considered the largest group able to be governed by the command of one man. So there were 100 soldiers under the command of a Roman centurion; 100 slave-soldiers under the command of a Mameluke emir; and, following the Roman model, there were 100 senators (two for each of fifty states) in the US Senate. More prosaically, 100 units comprise all the major currencies of the world – be they yuan, yen, dollars, euros, rials, rupees, dinars or pounds.

LET 100 FLOWERS BLOOM

There are repeated instances of 100 days of violent political and military activity, whether it be Napoleon's return in 1815 leading to the battle of Waterloo, Franklin D. Roosevelt's New Deal of 1933–36 or Robert Peel's first term in office.

工人阶级必须领导一切

Mao proclaims 'The 100 Flowers Campaign' in 1956.

But, for cold-blooded hypocrisy, few commands can quite equal Chairman Mao's unleashing of the first Cultural Revolution with the phrase 'Let a hundred flowers bloom' – an apparent liberalisation campaign that was reversed within months having been used to identify any dissident voice in order to destroy it.

HOMER'S CITY OF 100 GATES

Homer's chosen image for power was to describe Thebes, the capital of ancient Egypt, as a city of 100 gates; and from out of each one, at any moment, might pour 200 men riding chariots. Egyptian Thebes was known by its inhabitants as Waset. It should not be confused with Thebes in central Greece, a small but ancient Bronze Age city locked into an unprofitable rivalry with Athens and with its own numerical associations ever since Aeschylus wrote the play *Seven Against Thebes*.

99

Ar-Rahman (The Beneficent) ❋ Ar-Rahim (The Merciful)
❋ Al-Jabbar (The Compeller) ❋ Al-Latif (The Subtle) ❋
Al-Haq (The Truth) ❋ Al-Jame (The Gatherer)
❋ Al-Wakil (The Trustee) ❋ Al-Ghani (The Self-Sufficient)
❋ As–Sabur (The Patient) ❋ and 90 more

These are nine of the 'ninety-nine most beautiful names' that
are found inscribed in plaques around mosques, painted on
plates hung on bedroom walls, listed in ornamental scripts
and carved on marble, wood and plaster. Picked out in blue
and gold, they are printed on calendars given out by Muslim
businessmen and in the old days they were written on slips of
paper that could be dissolved in water and fed as medicine or
tied in a silver locket as a perpetual charm. They can also be
prescribed as spiritual and psychological cures. For instance,
Al-Ghani 'The Self-Sufficient' can be recited so that you will
become contented and not covetous; Al-Jame 'The Gatherer'
is often used in a Saint Jude-like manner for finding things
that have been lost; and Al-Wakil 'The Trustee' used to be
chanted by sailors when in danger on the sea.

The Koran seemingly makes a reference to this list (7:179) – 'To Him Belong the Most Beautiful Names' – though no number is actually given. The ninety-nine stands on the lesser authority of a hadith (saying of the Prophet): 'To God belong ninety-nine names'. The ninety-nine names are also a rosary of Muslim identity. Most Muslim names are formed from them, with the addition of the vital prefix of 'Abd' – that is to say, servant or slave of one of the names of God. They are a useful tool for a monotheistic culture to possess, for one of the essential dualities of all religious experience is the knowledge of unity but the need for diversity, and with a God of such vast power and mindless distance from humanity we need intercessors, be they Catholic saints,

A calligraphic rendering of the words 'Bismillah ar-rahman ar-rahim' (In the Name of God, Most Gracious, Most Merciful) in the shape of a hoopoe.

revered Sufi masters, Buddhist avatars, Shamanic healers or Judaic angels. The beautiful names of God encourage access from the faithful and promote an intimacy in prayer without deflecting from unity.

It is laughingly said that the camel with his inscrutable smile alone knows the 100th name, the hidden name, the Greatest Name, the name of power. It occurs somewhere in the Koran, though savants have searched through the holy text for centuries in vain. The verses al-Baqara, al-Imran and Taha are considered the most likely sources.

The old Christian communities of the desert also respected this number, for they read the word 'Amen' numerologically as $1 + 40 + 8 + 50 = 99$.

99 GOLD-UMBRELLA-BEARING RULERS

This wonderful image of the monarchy of old Burma comes from a reference to how Prince Limbin, having escaped the massacre of Mandalay palace in 1879, was able to style himself 'King of the Ninety and Nine Gold-Umbrella-Bearing Rulers'. It is a little like one of the old titles of the Byzantine emperors – the ruler who rules over those who rule – but in this case it seems to be quarried from popular tales from the Jataka, episodes from the imagined former lives of the Buddha.

95

95 THESES OF MARTIN LUTHER

Luther's Ninety-Five Theses were presented, as was the tradition of the time, by being nailed up on a door – at the Castle Church of Wittenberg on 31 October 1517. A German Franciscan friar, Luther's intent was to expose the greed, simony and errant nonsense of the sale of pardons from Hell fire by the Roman Catholic hierarchy. The pardons were being pushed in order to fund the building of Saint Peter's in Rome. The letter, modestly entitled 'On the power and efficacy of indulgences', would unleash the Europe-wide Reformation, though it began as a simple, almost respectful investigation of clerical abuse. It is not to be confused with the Thirty-Nine Articles of 1536 (a curious compromise which allowed the Church of England to be half-Calvinist, half-Catholic), nor the Forty-Two Very Protestant Articles of 1552, the Six Articles of 1539 or the Ten Articles of 1536. These were highly numerical decades for church politics.

88

---

88 – LUCKY IN CHINA

Eighty-eight is perceived to be a very good number in China, for '8' sounds similar to 'wealth' and 'profit' in Mandarin, while in the Cantonese dialect it sounds like 'fortune'. It is also the number of bright stars in the heavens. So 88 will make for a very popular top-floor penthouse-bar in a skyscraper, in a telephone number or on a car licence plate. It has also become a favourite phone text symbol for goodbye as '8–8' in Chinese sounds like 'bye-bye'. It's a superstition respected by the state, too. The Chinese chose to open the Olympic Games held in Beijing at 8pm on the 8th day of the 8th month of the year 2008.

The Chinese character for 8. Its shape suggests a person will have a great and broad future.

83

THE 83 *DÉPARTEMENTS* OF REVOLUTIONARY FRANCE

The many old provinces and *parlements* of France were swept away by the French Revolution. On 4 March 1791, France was divided into 83 new *département* units, named after unitary geographic units such as a valley or a mountain range. Each was to be governed by an official, the all-powerful *Intendant*, appointed by the central government. These officials were instructed to establish a departmental capital so that no area should be more than a day's horse ride from the head office. At the height of the conquests of Napoleon, this efficient bureaucratic Empire of Départements expanded to 130, but it later reverted down to 86. In 1860, the seizure of Nice and Savoy from Italy took the number up to 89, whilst victory in the First World War allowed for the Alsatian fortress of Belfort to become number 90. Further reorganisation and the addition of five overseas *départements* brought the number to its present tally of 101.

82

82 YEARS OF THE BUDDHA?

The most successful personality cult of all time is the Buddha, who died on Tuesday 15 May. The year of his death is much debated – 543, 486, 400 and 368 BC are mooted. However, his final words were faithfully recorded: 'Impermanent are all compounded things. Strive on heedfully.'

The Buddha was aged eighty-two (or eighty) when he died. He had lived for thirty one years as a prince, six years as an ascetic, and exercised his functions as a Buddha for forty-five years. Three months after his death, 500 enlightened persons gathered together in the Saptaparni hall at Rajagriha in central India. This was the First Buddhist Council, which collected together the sutra, the 'threads' of the Buddha's teaching. His body was cremated and the relic ashes placed in eight urns buried in earth mounds. A couple of hundred years later these would be excavated and the ash-filled urns subdivided to create the 84,000 memorial stupas erected by the Emperor Ashoka.

77

TWO SEVENS CLASH

Two Sevens Clash was the debut album from Culture, the roots reggae band led by Joseph Hill and produced in Kingston, Jamaica by Joe Gibbs. Its title refers to the date of 7.7.1977 – the day when 'two sevens met' – which the Rastafarian prophet Marcus Garvey predicted would be a day of chaos and apocalypse. As the liner notes of the album read: 'One day Joseph Hill had a vision, while riding a bus, of 1977 as a year of judgment – when two sevens clash – when past injustices would be avenged. Lyrics and melodies came into his head as he rode and thus was born the song "Two Sevens Clash" which became a massive hit in reggae circles both in Jamaica and abroad. The prophecies noted by the lyrics so profoundly

captured the imagination of the people that on July 7, 1977 – the day when sevens fully clashed (seventh day, seventh month, seventy-seventh year) a hush descended on Kingston; many people did not go outdoors, shops closed, an air of foreboding and expectation filled the city.'

77 NAMES OF THE GREAT HARE

Old Big Bum ❋ Frisky One ❋ Cat of the Wood ❋ Stag of the Stubble ❋ Swift as Wind ❋ Fidgety-Footed One ❋ Animal that No One Dare Name ❋ and 70 more

The extraordinary catalogue of seventy-seven names should be addressed to the Great Hare by lying down on the ground, placing your weapon on the ground and blessing him with a bent elbow and sincere devotion. If you are a hunter, you utter this prayer in praise of the hare so that you might bring him in dead. Or so the Welsh bard, Dafydd ap Gwilym (a contemporary of Chaucer's) advises us. Part flamboyant boast, part irritation (for a hare had just frightened off a girl that he was ardently courting in a reclusive grove), the prayer is enthused with the ancient hunter-gatherer's art of trapping the soul of an animal in a dream the night before the actual hunt. It is also larded with half-memories of the multiple titles with which you addressed the Great Hare as an aspect of the triple Mother goddess.

74

THE 74 HIDDEN NAMES OF RA

❈ The Becoming One ❈ Ra Of The Great Disk ❈
He of the Serene Face ❈ He Who Punishes with the Stake
❈ He Who Gives Light ❈ Lord Of Darkness ❈
The Flaming One ❈ The God Tefnut ❈ The Goddess Nut
❈ The Goddess Nephthys ❈ The Watery Abyss ❈
The Decomposed One ❈ Adu Fish ❈
The One of the Cat ❈ and 60 more

The seventy-four hidden names of Ra can be seen inscribed on the two rock pillars in the burial chamber of the House of Eternity built for Thutmoses III in the Valley of the Kings in Luxor. The divine names of Ra give explicit proof of the essential monotheism that coexisted with the bustling pantheons of deities within the ancient world. So, beyond the easy assimilation of Ra with other supreme deities of the sky and sun as worshipped and depicted in the great temple sanctuaries of Egypt (coupled with Amun, Ptah, Horus, Zeus-Ammon, Min, Khepri), there was an awareness that all forms of worship ultimately led back to a single source.

73

SETH'S 73 ACCOMPLICES

Although we do not know their names, the god Seth enlisted seventy-three accomplices when he tricked his brother-god Osiris. He enticed Osiris into coming to a feast, then, as an after-dinner game, the seventy-three joyfully took their turn in trying to fit into a cedar box. They all failed, for it had been manufactured to fit exactly the frame of Osiris, who – once he had entered – was held fast in a vice that allowed his brother to slam down the lid, seal the box and throw him into the Nile. There, he sailed out into the wide sea and was eventually washed ashore on the coast of Lebanon at Beirut.

73 BENEDICTINE RULES

The Benedictines follow a rule divided into seventy-three chapters of advice covering every detail of monastic order and the pursuit of a spiritual life. However, there remains an overriding stipulation that the rules exist to help and are not an end in themselves.

72

72 SHIA MARTYRS

Seventy-two has a key resonance in Islam as the number of followers who were killed alongside Imam Husayn at the massacre of Karbala on 10th October 680 (tenth of Muharram). It is the central sacrificial tragedy of the Shia community and a narrative that – for all their doctrinal differences – is undisputed by the Sunni majority, for all Muslims honour the story of how the brave grandson of the Prophet responded to a call for help from the people of southern Iraq to come to their aid and protect them from the abusive government which had taken control of the Arab Empire of Faith and converted it into a corrupt monarchy. When Husayn arrived, the people turned their back on him, and his small band of followers and relatives were shunned, then progressively starved of food and water before being murdered in a shower of arrows by an army of 4,000. It is in memory of this fearful day, when the Muslim community deserted their natural leader, that the Shia launch into an emotional annual festival of self-examination and contrition known as Ashura, 'the tenth'.

72 CATHAR BISHOPS

The Cathar church was ascetic and dualist – believing in both a good and evil God. This church is thought to have been governed by a council of seventy-two bishops (episcopi), drawn from 360 priests (mahistag), assisted by the Elect and the Hearers of their communities. The council of bishops selected twelve Apostles (hamozag), who chose one of their number to be leader (achegos). These heretic communities – known variously as Manicheans, Paulicians and Bogomils – were ruthlessly persecuted out of existence by Catholic and Orthodox church authorities, most famously in southern France by the thirteenth-century Albigensian Crusade.

The number chosen by Cathars may have had links with one or other earlier beliefs: that there were seventy-two disciples of Christ who preached the truth to the world in seventy-two different languages; that King David could sing the psalms to his lord in seventy-two different ways; that the sanctuary-cathedral of the Holy Grail had seventy-two side chapels; and that Adam and Eve had seventy-two children, in thirty-six sets of twins.

72 SACRIFICES TO ODIN

In the eleventh century, Adam of Breven stumbled across evidence of one of the last great pagan cult shrines in Europe, at Uppsala in Sweden, while researching his history of north Germany and Scandinavia. Whether in his imagination or in reality, he reported that at one great festival, seventy-two sacrificial victims were hung to Odin (Wotan/ Woden) in a sacred grove of yew trees. In another description he suggests that nine male sacrifices (animals as well as humans) were made a day. If this was part of an eight-day festival, it may be one of the root sources of the power of

Odin, the one-eyed god of prophesy and foreknowlege.

seventy-two, for he also mentions that the shrine was dominated by a powerful trinity of gods which, seen in their own triple aspects, makes nine.

Adam wrote: 'In this temple, entirely decked out in gold, the people worship the statues of three gods. Thor occupies a throne in the middle of the chamber; Wotan and Frikko have places on either side ... Thor, they say, presides over the air, which governs the thunder and lightning, the winds and rains, fair weather crops. The other, Wotan, that is, the Furious, carries on war and imparts to man strength against his enemies. The third is Frikko, who bestows peace and pleasure on mortals. His likeness, too, they fashion with an immense phallus ... For all their gods there are appointed priests to offer sacrifices for the people. If plague and famine threaten, a libation is poured to the idol Thor; if war, to Wotan; if marriages are to be celebrated, to Frikko.'

70

―――――

Jebus ❀ Zion ❀ Shalem ❀ Urushalem ❀ City of David ❀
❀ Yerushalayim ❀ Ariel ❀ Aelia Capitolina ❀ God's City ❀
Al-Quds/El-Kuds (Holy City) ❀ Faithful City ❀ Eternal
City ❀ City of Peace ❀ Beautiful City ❀ Blessed City ❀
Ir Ha-Kodesh (City of Holiness) ❀ Terra Sancta
❀ Bir Aruakh (well of souls) ❀ and 62 more

Jerusalem is the city of seventy names according to Rabbinic tradition. Those listed above don't quite reach that number but when translated back into Arabic, Syriac, Aramaic, Hebrew, Greek and Latin, and listed with variant spellings and vulgarisations by other languages, they soon add up. The name 'Jerusalem' itself has its origins as the threshing place and place of sacrifice of the Jebusites, just outside their citadel of Shalem, dedicated to the god Uru – hence 'Urushalem'.

A fateful connection between Jerusalem and the number 70 is recalled by its capture by the Roman army in 70 AD, who slaughtered the population. The Old Testament also records that seventy men accompanied Moses up to Mount Sinai, that the Babylonian captivity lasted seventy years, that Moses was mourned for seventy days after his death and, most famously, in Psalms 90:10, that 'Our Life lasts three-score years and ten.'

70 CUPS OF POISON

The Seventy Cups of Poison are the various sorts of drugs, drinks, devilment and debaucheries available to man. The phrase appears in a description of a seventeenth-century parade of the guilds of Istanbul: 'comics, mimics and mischievous boys of the town, who have exhausted seventy cups of the poison of life and misrule, crowd together day and night ... they are divided into twelve companies, the first gypsy, the last Jewish, which included two hundred youths all tumblers, jugglers, fire-eaters, ball-players and cup-bearers.'

70 HOLY IDIOTS OF SUFISM

We are not permitted to know the names of the Holy Idiots, for, according to Sufi tradition, 'My saints are all under the vaults of Heaven, nobody knows them but I.' But the story is that it is the Holy Idiots (only seventy of whom are alive at any one time) who spread divine love amongst mankind. Some are noble, some imbecile, some holy, some coarse and some pious. The Sufi also like to remember that even the Prophet felt the need to pray for forgiveness seventy times a day; that the Prophet recited the Koran seventy times on his mystical journey through the heavens; and that 70,000 veils of perception stand between God and the believer.

The historian John Freely, resident in Istanbul in the 1950s, believed that he knew the identity of one of these seventy fools, who had his station on the bridge over the Golden Horn, where he would greet all and sundry. He was known as the lunatic with horns, for he was decorated with the horns of billy-goats, gazelles and rams, and he never forgot a name, even recognising people twenty years after he had last seen them. He was last seen setting off to fetch some new horns, taking the long road to Abyssinia.

THE SEPTUAGINT

The Septuagint is the name for the Greek translation of the Hebrew Testament made in Alexandria in Egypt in the fourth century BC. Believed to be either a miraculous harmony of seventy scholars working separately to produce an identical textual translation, or a body of seventy scholars working together to produce a single agreed text – which is arguably an even more miraculous occurrence.

GERALD OF CREMONA'S BOOK OF SEVENTY

The *Book of Seventy* is a remarkable medieval encyclopedia of science, compiled around 1187 by Gerald of Cremona, who created it by translating books attributed to Geber (Jabir ibn Hayyan), an eighth-century Perisan polymath. Gerald himself was a prolific scholar, based mainly in the newly Christianised city of Toledo in Spain, at its heyday as a free-thinking intellectual powerhouse. His book included sections on algebra, trigonometry, optics, medicine, astronomy and chemistry, or alchemy, as it was then known).

66

66: THE NUMBER OF ALLAH

Sixty-six is the number derived from 'Allah' through cryptic numerology (giving letters certain numerical values). This value may be referred to in Islamic architecture when a dome or an arch is encircled in sixty-six floral rosettes. The science of cryptic numerology is believed to have been perfected by Imam Ali, the prophet Muhammad's cousin, first disciple and son-in-law. Other key numbers extracted from the names of God include 289, which is derived both from 'Al-Rahman' and 'Allaha Akbar' (God is Great), and 256, from 'Nur' (The Light).

Sixty-six can also be separated out by a Shia-inclined mystic to 6 and 6, which becomes 12, the number of the true Imams of Shiite Islam.

A North African magic square showing the numerical value derived from the letters of 'Allah'.

64

THE KAMASUTRA'S 64 ARTS OF LOVE

In the Kamasutra, the Sanskrit guide to the arts of love, there are sixty-four varieties of sexual acts.

The manual was composed in India in the third century AD by a Hindu Vedic philosopher called Vatsyayana, though the root origin of its collected wisdom is said to have been Nandi, the sacred bull, listening to the bed play of the divine couple, Shiva and Parvati. The book's advice, which should not be confused with tantric sex (largely an invention of the 1960s), ranges widely, discussing how best to achieve a virtuous family life. But it is the sex advice which has kept the book in circulation for 2,000 years. This comprises discourses on stimulating desire, embracing, caressing and kissing, marking with nails, biting and marking with teeth, copulation positions, slapping by hand (and moaning), virile behaviour in women, oral sex, foreplay and 'conclusions'.

The Kamasutra was first published in English in 1883, under the direction of the polymath and explorer Richard Burton. It was soon filleted down and produced with explicit Indian illustrations. Modern editions include a pop-up version.

60

---

60 DEGREES OF SUMERIA

Sixty is the base number of the Sumerian number system, fully evolved in Mesopotamia (modern-day Iraq) by 3000 BC, and it remains the essence of how we measure time: sixty seconds in a minute, sixty minutes in an hour. The number is also the base of the 360 degrees of a circle, as in the fully imagined sky of the Sumerians (of which only a portion was ever visible from temple roofs), divided into six houses of 60 degrees. In Sumerian culture, the number 1 was expressed by a simple wedge, cut into clay or wood, and 60 by a great wedge.

Sixty has the versatility of being neatly divisible by 30, 20, 15, 12, 10, 6, 5, 4, 3 and 2, and therefore makes for easy subdivision of irrigated land and the harvested crops which were initially gathered in sixty-fold sheaves, just as in pre-decimal English currency sixty pennies (60d) were a crown (five shillings/5s).

59

The fifty-nine regicides were those members of a hand-picked tribunal of over a hundred Parliamentarians who had the courage of their convictions and placed their signatures and seals on the death warrant of King Charles I in 1649. Nineteen members of the Tribunal refused to sign the warrant and a number of those requested to serve on it, such as all twelve of the High Court judges summoned, refused to attend.

The regicides would have become a foundation senate, placed in a national pantheon, if England had remained a republican commonwealth. However after the Restoration of Charles II, the Royal Pardon specifically excluded all the fifty-nine regicides; some were tried and pardoned, others executed or imprisoned for life in the Tower. Others escaped into exile (including three who found refuge in Connecticut), while some of those already dead – most famously Oliver Cromwell – were exhumed and executed as traitors, and their heads displayed outside Westminster Hall, where they had sat in judgement on their king.

56

56 PILLARS

In prehistoric Britain, fifty-six stone pillars stood in the outer circle of Stonehenge. In more recent times, the National War Memorial in Washington, erected after World War II, commemorates the dead with fifty-six pillars (also the number of signatures on the 1776 Declaration of Independence of the thirteen states). And in Beijing's Tiananmen Square, to commemorate the sixtieth anniversary of the People's Republic, fifty-six towering red columns were erected to represent all the 'equal, united and harmonious' ethnic groups of China.

2nd Circle.

Altar.

52

---

52 SHAKTI PITHAS

Shakti, the great Mother Goddess of Hindu India, tore herself into pieces in order to get her consort, Shiva, to halt his dance of the destruction of the universe. Shiva became locked into a trance of grief and the world was saved. This self-immolation scattered fifty-two pieces of the great goddess, which fell from the sky of the heavens. Not all survived, or can be recognised, but four have become sites of great respect: two in Orissa, one in Assam and one in West Bengal. There are other lists that prescribe 108 sites, another powerfully significant number in the East.

THE MAYAN CALENDAR'S 52-YEAR CYCLE

The Mayans' fifty-two-year cycle is created by observing how the combination of their two simultaneous calendars – the 260-day-long Tzolk'in fertility calendar and 365-day-long Haab solar year – fitted into a naturally repeating cycle over a fifty-two year time span.

50

PENTECOST

The Pentecost is the most magical instance of a feast based on a fifty-day unit of time. And although it is now considered to be a great day of spiritual transformation (celebrated by Christians as Whitsun – the day the Holy Spirit came down upon the Apostles – and by the Jews as the day Moses brought the Law of God to them), in origin it was a simple harvest festival. It was held fifty days after Passover/Easter, a day when the tithe would be assessed in the Day of First Fruits (an ancient tax assessment date) alongside prayers of thanks and celebration to mark the Feast of Weeks/Festival of Reaping.

OLD TESTAMENT UNITS OF 50

Units of fifty appear throughout the Old Testament. In Genesis (18:23) we hear that Abraham wished to persuade the Lord to spare the city of Sodom, if he could but identify fifty good men. In Numbers it is related how a tithe of one from every fifty (be it man or beast, cattle, asses and sheep) be given over to the Levites in charge of the Lord's Tabernacle. Noah's Ark was fifty cubits in breadth. And there were fifty golden claps on the hangings of the Tabernacle.

However, fifty was perhaps not so significant a number. In the Elamite and Akkadian languages of ancient Babylon, it seems to have encompassed the concept of 'all'. So when we hear that during one of the adventures of Gilgamesh he was accompanied by fifty companions, it might just be a body of men. Likewise in the Bible (II Kings 1:9), when Elijah faces the King of Samaria's 'captain of fifty with his fifty', a group consumed by a fire from Heaven, as are its reinforcements, 'another captain of fifty with his fifty'.

50 ARGONAUTS

Jason ✳ Orpheus (the lyre-playing musician) ✳ Mopsus the seer ✳ Heracles and his male love of the moment, the handsome young Hylas (who gets kidnapped by water nymphs) ✳ Pollux the champion boxer who kills the king of the Bebrycians ✳ Shape-shifting Periclymenus ✳ Fast-footed Euphemus ✳ Winged Calais and Zetes (sons of the North Wind who repel the Harpies) ✳ and 40 more

The Argo, which had a magical keel crafted out of a sacred oak from the oracle of Dodona, was crewed by fifty heroes of ancient Greece – the Argonauts. Jason was the leader of this warrior band (sometimes referred to as the 'Minyans') sent on what was presumed to be a suicidal quest by King Pelias, his usurping half-uncle. Their mission was to sail to Colchis (Georgia) and seize possession of the Golden Fleece of a divine ram that hung from a tree in a grove sacred to Ares, god of war, guarded by a sleepless dragon.

Every city in Greece liked to imagine that they contributed a hero to this mythical band, which means that the list has

Jason and the Argonauts, still getting into scrapes in videoworld.

had to grow in number, though if you examine the text of Apollonius of Rhodes, written in third-century Alexandria, it is easy enough identify all the named Argonauts. Even this cast, however, numbers fifty-five, though by juggling who comes on, as others go off, the good ship Argo, it is just about possible to keep to fifty.

If you add other famous names and such ubiquitous heroes as Bellerophon, Nestor, Perseus, Atalanta and Theseus, you can grow the crew to eighty, which has a hidden harmony with the text of Apollonius, who has embedded eighty *aitia* in his epic. These are short verse sequences which give the mythical origins of such curious things as the sacred water-carrying race held on the island of Aegina or how the island of Thira is linked with Libya. The final text comprised 6,000 lines, which can be recited in one day to a reasonably alert ancient theatrical audience.

49

SEVEN SEVENS ARE 49

Forty-nine alone escapes the Eastern suspicion of anything to do with the number 4 (which has a tonal connection with the Chinese word for death). This is because it is the sum of seven times seven, and 'seven' is very propitious because it sounds like 'arise' and also can mean 'togetherness'. For the superstitious, rather than writing forty-nine by itself, seven times seven is often used or tacked on beside it. So forty-nine has become the Eastern world's preferred length of time for fasting and cleansing rituals, as well as being the period of time for a requiem ritual after a death.

49 TITLES OF THE BLESSED VIRGIN MARY

Mother of God ✣ Mother most chaste
✣ Mother without stain ✣ Mirror of justice ✣ Rose mystic
✣ Star of the morning ✣ Queen of peace✣ and 42 more

Here are seven of the forty-nine titles of Mary, the Jewish mother of Jesus. She was believed to have been 49 years old when her son ascended into Heaven. Mary herself died either eleven years after the death of Jesus (that is, in 41 AD), or twenty-four years after the Ascension, when she physically ascended to Heaven.

Most of the earliest Christian veneration of the Virgin Mary originated from Egypt, which gave us the first hymns in her honour, such as 'Beneath Thy Protection', created decades before the official triumph of Christianity. The early Church father, Origen of Alexandria, is credited with awarding Mary her familar Orthodox title of Theotokos ('God Bearer'). Origen himself was later dismissed as a heretic, due to his belief in the pre-existence of souls, the final reconciliation of all creatures, including the devil, and the subordination of the Son of God to God the Father.

Orthodox Marian icons from the eighteenth century Russia.

42

THE KABBALAH'S 42-LETTERED GOD

The Kabbalistic traditions of mystical Judaism give tremendous respect to the hidden forty-two-lettered name of God, as well as the numerical values observed to be within the Hebrew word *Atziluth* ('Emanation'), which was believed to be one of the most creative and powerful appellations of Jehovah. This respect is echoed in the Gospel of Saint Matthew, which gives forty-two links in the genealogy of Christ (from Abraham).

42 – LIFE, THE UNIVERSE AND EVERYTHING

In Douglas Adams' *Hitchhiker's Guide to the Galaxy*, the computer Deep Thought takes 7.5 million years to work out that the Answer to the Ultimate Question of Life, the Universe and Everything is '42' – even if in the process the question had been forgotten. It is an answer that must disconcert Japanese readers, for 42 in Japan is like 49 in Chinese: when pronounced 'four' and 'two', it sounds horribly similar to 'unto death'.

42 ASSESSORS

The Judgement of the Dead in ancient Egypt revolved around forty-two, the number of the little, seated gods of judgement, the Assessors, depicted in innumerable tombs and in the illustrated scrolls of the Book of the Dead.

They look somewhat formulaic and innocuous, lacking in character and symbols of identity, but it is clear from surviving texts that they were feared as a very precise tribunal of inquiry looking in great detail at every infringement of the deceased. The high moral nature of the ancient world is not always acknowledged, but can be most emphatically felt by reading your way through the forty-two confessions that a mortal soul was expected to be able to make (part of a Declaration of Innocence by the deceased in the Hall of Justice and the broad hallway of the Two Truths), alongside the more actively positive assertions. There was no absolute formula of the confessions to be made by each individual, but more an imagination of what was right for each person seeking to live a just existence.

The Weighing of the Heart ritual, shown in the Book of the Dead of Sesostriss; a group of Assessors are shown, seated, along the top.

The list below – the Forty-Two Confessions of Maat – is taken from the Ani papyrus of the Book of the Dead as translated by Wallace Budge, the British Egyptologist and anthropologist whose ghost is said to still haunt the British Museum.

I have not committed robbery with violence ✸ I have not stolen ✸ I have not slain men and women ✸ I have not stolen grain ✸ I have not purloined offerings ✸ I have not stolen the property of the god ✸ I have not uttered lies ✸ I have not carried away food ✸ I have not uttered curses ✸ I have not committed adultery, nor have I lain with men ✸ I have made none to weep ✸ I have not eaten the heart – by pretending remorse that I did not feel ✸ I have not attacked any man ✸ I am not a man of deceit ✸ I have not stolen cultivated land ✸ I have not been an eavesdropper ✸ I have not slandered ✸ I have not been angry without just cause ✸ I have not debauched any man ✸ I have not debauched the wife of any man ✸ I have not polluted myself ✸ I have terrorised none ✸ I have not transgressed the Law ✸ I have not been wroth ✸ I have not shut my ears to the words of truth ✸ I have not blasphemed ✸ I am not a man of violence ✸ I am not a stirrer up of strife or a disturber of the peace ✸ I have not judged anyone with undue haste ✸ I have not pried into matters that do not concern me ✸ I have not multiplied my words in speaking ✸ I have wronged noone ✸ I have done no evil ✸ I have not worked witchcraft against the King ✸ I have never stopped the flow of water for the crops ✸ I have never raised my voice and spoken in anger ✸ I have not blasphemed God ✸ I have not acted with arrogance ✸ I have not stolen the bread of the gods ✸ I have not carried away the khenfu cakes left out for the Spirits of the dead ✸ I have not snatched away the bread of the child, nor treated with contempt the god of my city ✸ I have not slain the cattle belonging to the god or treated with contempt the god of my city ✸

40

===

40 DAYS AND 40 NIGHTS

Forty days and forty nights is a powerful and emblematic unit of time. It is the prescribed length of time when the earth was flooded in both the biblical and Babylonian epics. The Jews wandered for forty years in the wilderness before reaching the Promised Land; Moses spent forty days on the mountain with his God before returning with the Ten Commandments; Elijah also spent forty days meditating on Mount Horeb, just as Jesus spent forty days and forty nights meditating in the wilderness of the Jordanian desert after his baptism. Jesus spent forty hours in the tomb, and after his resurrection he would spend just forty days on earth; Jerusalem would be destroyed by the Romans forty years after the crucifixion of Jesus; and the Prophet Muhammad would receive his first revelation (whilst meditating on a mountain top) aged 40. Forty is the prescribed length of time for the Lenten fast and a paschal taper should be kept burning for forty days between the festivals of Easter and Ascension Day. There are forty large stone pillars at Stonehenge.

What could be behind the respect for this number?

Some have suggested that it based on the number of fingers and toes possessed by a pair of lovers, or by a child looking at the universe of possibility offered up by the forty fingers and toes of its two parents. But it is also the number of days that the Pleiades constellation disappears from view, which the Babylonians anciently associated with the rainy season. Another potent idea is that respect for 40 is based on the weeks of a human pregnancy, which seems to have been mirrored in the number of days that many cultures thought was required for the soul to leave the body, and the period in which you could bury a corpse (or leave it out for the birds and the foxes) before gathering the bones for ceremonious reburial. Still today, in the Islamic and Orthodox worlds, the fortieth day after death is celebrated by a gathering of the extended family at the tomb, almost as if they are empowering the departed soul to take the final journey.

Jesus, tempted by devils during his forty days and nights in the wilderness, as depicted on a mosaic at Saint Mark's Cathedral, Venice.

THE SPLITTING OF A HAIR INTO 40 PARTS

The splitting of a hair into forty parts was believed in the magically inclined medieval times to have been achieved by the six great physicians of antiquity – Plato, Hippocrates, Socrates, Aristotle, Pythagoras and Galen. The physicians then used it to make a ladder in which science could ascend to the heavens, but there they failed to find a cure for death and returned to earth. Sometimes their number is extended by allowing King Philip II of Macedon to join this band.

THE *ARBA'IN* – THE 40

In the Sufi traditions of Islam, a reference to the *arba'in* comes with a wealth of potent associations. It can be the number of Ali's disciples (from which all Sufi brotherhoods trace their spiritual descent) and it can be a reference to the forty living Sufi saints who are felt to be present on earth at any one age to help keep humans on the right path. It is also an act of scholarly piety to collect together the forty most personally relevant hadith (sayings of the Prophet) from the thousands that have been assembled. The number 40 is also embedded in the great foundation stone of Sufi mystical teaching, al-Ghazali's *Revival of Religous Sciences*, which is composed of forty chapters that prepare believers for their deathbed. A fortieth of your wealth is also the customary assessment of the *zakat* – the tithe donated by each believer to allow the community to care for the sick, the poor and the elderly.

On a less spiritual plain, within the Muslim world the tale of Ali Baba and the forty thieves is so well-known, and the police so universally despised for their corruption, that any reference to 'Ali Baba' is taken to mean 'Watch out – there are police about.'

39

THE 39 STEPS

John Buchan's 1915 action novel has been filmed at regular intervals, following Alfred Hitchcock's classic version with Robert Donat as buccaneering hero James Hannay. Buchan was famously inspired to write the novel by his daugher counting the stairs of the nursing home in Broadstairs, where he was convalescing. He turned the phrase into a key mystery of the novel, and Hannay's eventual discovery of its meaning (it is the number of steps down a cliff path to a waiting yacht) helps keep Britain's military secrets intact from the Germans.

Hitchcock significantly changed Buchan's plot for his 1935 movie, writing a climactic music hall scene in which 'Mr Memory' is asked 'What are the 39 Steps?' and is about to reveal the answer ('The 39 Steps is an organisation of spies, collecting information on behalf of the foreign office of ...') when he is shot dead. An equally evocative twist was introduced in the 1978 film starring Robert Powell, where the thirty-nine steps turn out to be the number of stairs in the clock tower of Big Ben.

33

33 – NUMBER OF COMPLETION

Thirty-three is an ancient number of completion: the age when Christ was crucified; the years in which King David reigned. It also marks the number of divinities in the public festivals of the official calendar of the Persian Empire, and in the Hindu tradition three sets of eleven deities appear frequently as an auspicious pantheon of thirty-three. In Muslim tradition the ninety-nine beautiful names of God are recited with rosaries made from thirty-three prayer beads each used thrice, while the Hizb al-Wiqaya is a prayer of personal protection collected from thirty-three verses that invoke Koranic protection and divine names.

In broader cultural contexts, the number was chosen by Dante to structure his *Divine Comedy* (composed of three sets of thirty-three chapters); it expresses the number of spiritual ranks within Freemasonry, and the blows with which Shakespeare records death being delivered to Julius Caesar ('When think you that the sword goes up again? Never, till Caesar's three and thirty wounds be well avenged').

32

32 SIGNS OF THE UNIVERSAL RULER

Well-set off feet ❋ Eight lucky symbols on soles of feet
❋ Projecting heels ❋ Long fingers ❋ Soft hands and
feet ❋ Netted hands and feet ❋ Prominent ankles ❋
Antelope limbs ❋ When standing, hands reach to knees
❋ Uncircumcised penis ❋ Golden colour ❋ Soft skin ❋
One hair to each pore of skin ❋ Hairs of the body are
black, rise straight and curl to the right ❋ Straight of body
❋ Seven prominences ❋ Front part of body like a lion ❋
Space between the shoulders filled out ❋ Height equal to
outstretched arms ❋ Even shoulders ❋ Keen taste ❋ Lion
jaw ❋ Forty teeth ❋ Even teeth ❋ Not gap-toothed ❋
White teeth ❋ Large tongue ❋ Voice like Brahma and soft
as a cuckoo's ❋ Black eyes ❋ Eyelashes like an ox ❋ White
hair between the eyebrows ❋ Head is the shape of a cap

These are believed to be the thirty-two bodily marks of the
Buddha and by inference the signs needed to be found in his
next incarnation. Although the interpretation of such signs
was condemned, they appeared in all popular forms of the
legend of the Buddha.

To add a further touch of complexity there are an additional eighty minor marks, plus eight lucky symbols on the Buddha's foot: Wheel of the Law, Conch SHell, Umbrella, Canopy, Lotus, Vase, Pair of Fish and Mystic Knot. This group of symbols is often used to decorate prayer flags and devotional books, and as motifs in Buddhist shrines.

In Islamic tradition the redeemer of mankind who comes at the end of the world is the Mahdi, about whom very little is known from the approved texts, though this has not stopped a populist interest in the 'signs'. The Mahdi's physical appearance is based on the remembered physique of the Prophet, with 'a face that shines like the moon, a high forehead and a long, thin, curved nose. He will fill the earth with fairness and justice as it was filled with oppression and injustice, and he will rule for seven years.'

32 GRAINS OF AN ENGLISH COIN

Thirty-two grains of English wheat, taken from the middle of the ear of corn (so as to confound cheats and counterfeiters) was the official weight of an English silver penny according to the reforms of old King Offa of Mercia (757–796), undertaken in parallel to those of the Emperor Charlemagne in mainland Europe. Twenty of these pennies should weigh in at an ounce (to give the equal of the old Latin *solidus* coin of the Romans and the English shilling) , and twelve such ounces produced the royally approved standard of a Tower Pound, worth 240 silver pennies. All of which said, in 1284 King Edward I switched the currency off the wheat standard back to the barley grain.

31

THE GREEK LEAGUE OF 31 AGAINST PERSIA

Athens ❀ Corinth ❀ Megara ❀ Sicyon ❀ Tegea ❀ Phlius
❀ Troezen ❀ Anactorium ❀ Epidaurus ❀ Orchomenos ❀
Chalcidice ❀ Mycenae ❀ Tiryns ❀ Hermione ❀ Eretria
❀ Styra ❀ Plataea ❀ Aegina ❀ Ambracia ❀ Potidaea ❀
Cephalonia ❀ Lepreum ❀ and nine other cities whose
contingents numbered under 200 men

This league of thirty-one Greek cities defeated the army of
the Persian Empire at the Battle of Plataea in 479 BC, the
year after Thermopylae and the sack of Athens. The Greek
military victory was the gift of the Spartans, who both led
and numerically dominated the army. A victory monument
was created by melting down some of the captured bronze
armour of the enemy dead, which was shaped into a triple
serpent and presented as an offering to the oracle at Delphi.
It was later moved by the Emperor Constantine to embellish
his new imperial capital of Istanbul and set in the middle of
the Circus Track. There it remains, and those with sharp eyes
may pick out some of the civic inscriptions that are just still
legible on this monument.

30

30 DAYS OF RAMADAN

A Muslim should fast for the thirty days of the month of Ramadan to commemorate the first revelation of the Koran. The fast lasts from sunrise to sunset, avoiding food, drink, sex and cigarettes. After the evening prayers it is traditional for a pious Muslim to listen to a recitation of the Koran, which is divided into thirty equal sections so that the Holy Book can be heard in its entirety during this month. Within the month there is one night (the 21st, 23rd, 25th, 27th and 29th are favoured) in which the gates of Heaven are opened to the prayers of the believers, the Laylat al-Qadr, the night of power, where prayers and petitions are traditionally believed to be more efficacious than 1,000 months of dutiful piety.

30 PIECES OF SILVER

In the Christian tradition 30 is associated with betrayal. Saint Matthew's Gospel tells the story of the thirty pieces of silver that Judas was paid to reveal his master with a kiss as he addressed him in the Garden of Gethsemane so that the temple guards of Jerusalem could identify and arrest him. Later, in remorse, Judas attempts to return the coins before hanging himself, but the chief priests, knowing it is now dirty blood money, refuse to place it in the temple treasury but instead use it to buy a potter's field as a burial place for penniless foreigners. Thirty clearly had a textual relevance for Saint Matthew, who knew that in the Old Testament book of Exodus it was the price of a slave and that in Zachariah it was the price put on his labours which he in his anger gives to a potter.

Christian artists usually have the pieces of silver in a bag but those who have sought to illustrate them usually depict the Tyrian shekel. A half-shekel was the accepted form of annual tithe paid by every adult Jewish male to the temple. It bore the head of Melkarth (Phoenician Hercules), the hero-god of the city of Tyre, and was widely esteemed for the purity of its silver. In medieval times, in order to avoid the use of this unlucky 30, and to request less than that which was paid for the Son of God, the ransom demanded for the release of a captured king would always be set at a maximum of twenty-nine units.

The Tyrian half-shekel – as received by Judas Iscariot.

29

THE CANTERBURY TALES' 29 PILGRIMS

Chaucer's tale-tellers: Knight ❋ Miller ❋ Reeve ❋ Cook ❋ Man of Law ❋ Wife of Bath ❋ Friar ❋ Summoner ❋ Clerk ❋ Merchant ❋ Squire ❋ Franklin ❋ Physician ❋ Pardoner ❋ Shipman ❋ Prioress ❋ Monk ❋ Nun's Priest ❋ Second Nun ❋ Canon's Yeoman ❋ Manciple ❋ Parson ❋ Narrator

And those who don't tell tales: Host ❋ Plowman ❋ Yeoman ❋ Canon ❋ Second Priest ❋ Third Priest and Five Guildsmen (Haberdasher, Carpenter, Weaver, Dyer, Arras-Maker)

Chaucer tells us that there are 'well nyne and twenty' pilgrims in the company that sets off from Southwark to visit the shrine of Saint Thomas-à-Becket in Canterbury. But once you start list-making you find that such numerical certainty proves evasive, for there are thirty-four identifiable characters in his text, of whom twenty-three tell a tale. I like to imagine that the Host and the Five Guildsmen would have been made to perform if Chaucer had lived long enough, for *The Canterbury Tales* was almost certainly a work in progress, which Chaucer happily tinkered with all his life.

25

Adam ✳ Idris (Enoch) ✳ Nuh (Noah) ✳ Hud ✳ Saleh ✳
Ibrahim (Abraham) ✳ Isma'il (Ishmael) ✳ Ishaq (Isaac) ✳
Lut (Lot) ✳ Ya'qub (Jacob) ✳ Yousef (Joseph) ✳ Shu'aib ✳
Ayyub (Job) ✳ Musa (Moses) ✳ Harun (Aaron) ✳ Dhu'l-
kifl (Ezekiel) ✳ Dawud (David) ✳ Sulaiman (Solomon) ✳
Ilias (Elias) ✳ Al-Yasa (Elisha) ✳ Yunus (Jonah) ✳ Zakariyya
(Zachariah) ✳ Yahya (John) ✳ Isa (Jesus) ✳ Muhammad

Though the Koran acknowledges many thousands of
Prophets, only twenty-five are specifically mentioned in
the Koran, though at times there may seem to be many
more due to the variable use of honorifics such as Hazrat,
Si, Sidi or Moulay – which very roughly translate into
English as 'His Honour', 'Sir,' 'My Master' and 'My Lord'.
These prophets also provide a list of popular given names
within the Islamic world.

25 WARDS OF THE CITY OF LONDON.

Aldersgate ❋ Aldgate ❋ Bassishaw ❋ Billingsgate ❋
Bishopsgate ❋ Bread Street ❋ Bridge and Bridge Without
❋ Broad Street ❋ Candlewick ❋ Castle Baynard ❋ Cheap
❋ Coleman Street ❋ Cordwainer ❋ Cornhill ❋ Cripplegate
❋ Dowgate ❋ Farringdon Within ❋ Farringdon Without ❋
Langbourn ❋ Lime Street ❋ Portsoken ❋ Queenhithe
❋ Tower ❋ Vintry ❋ Walbrook

There have been twenty-five wards of the City of London
for the last 1,000 years. They occasionally get bumped up by
a sub-division, or down by an amalgamation, but happily we
are set on twenty-five at the moment. In ancient days these
wards allowed for a mosaic of parish-like administration, little self-governing communities with their own assemblies
(*wardmote*), wells, local markets, cemeteries, systems of public
order (three elected beadles), and charities presided over by
an Alderman who formed a sort of Senate of London, the
Court of Aldermen. From this Court, the separate system
of Livery Companies (trade guilds) elected a Lord Mayor,
replaced every year to soften any authoritarian tendencies.

24

24 TIRTHANKARAS OF THE JAIN

Lord Rishabha/Adinath ☀ Lord Ajitnath ☀
Lord Sambhavanath ☀ Lord Abhinandanath ☀
Lord Sumatinath ☀ Lord Padmaprabha ☀ Lord Suparshvanath
☀ Lord Chandraprabha ☀ Lord Suvidhinatha ☀
Lord Shatalnath ☀ Lord Shreyanshanath ☀ Lord Vasupujya ☀
Lord Vimalnath ☀ Lord Anantnath ☀ Lord Dharmanath ☀
Lord Shantinath ☀ Lord Kunthunath ☀ Lord Aranath
☀ Lord Mallinath ☀ Lord Munisuvrata
☀ Lord Naminatha ☀ Lord Neminatha
☀ Lord Parshwanatha ☀ Mahavira

These are the succession of Jain philosopher-prophets, who conquered both love and hatred and so were able 'to show the true way across the troubled ocean of life'. The line begins with Shri Rishabha and reached its penultimate exponent in Shri Parshwananatha who was active in the ninth century BC. The ultimate Tirthankara, Mahavira, was the senior contemporary and near neighbour of the Lord Buddha in the sixth century BC.

24 NUBA

Whether in poetic mythology or practice, the poet and musician Ibn Ziryab (789–857) is believed to have brought the corpus of traditional music from the court of the Abbasid Caliphate in Baghdad to Muslim Spain, from where it filtered out to the troubadors of southern Europe and the classical musical traditions of North Africa and the Near East. His repertoire of symphonic music established the twenty-four modes or Nubas – one appropriate to each hour of the day, and themselves divided into five parts – though only a fraction of this heritage survives today. Ziryab is also credited with adding the fifth string to the oud and of colouring the previous four strings with the symbolism of the four humours, which (when in balance) supported the creation of the worthy soul.

Ibn Ziryab entertaining the court at Cordoba with his new-fangled, five-stringed oud.

Ibn Ziryab is variously claimed to have been of Iranian or Kurdish background mixed with African blood, for the name 'Ziryab' means 'blackbird', which was both a comment on his voice and the colour of his skin. He was expelled from the Abbasid court due to the jealousy of his old master, after which he travelled across all the lands of Islam before settling among the rival Umayyad dynasty who ruled from Cordoba in southern Spain. Here he was honoured with a salaried position at court and in return gave freely of his immense knowledge – be it on music, cooking, health, medicine or courtly manners or deportment.

24 –THE AGE AT WHICH THE PROPHET MANI WAS CHOSEN

The Prophet Mani was the founder of Manichaeism, a dualist religion that fused elements of Christianity, Zarathustrianism and Buddhism. Mani taught that good and evil were locked in eternal (Manichean) combat, but that salvation could be achieved by self-denial, fasting and chastity. Born in Ctesiphon in modern-day Iraq, around 216 AD, Mani was chosen to be a prophet at the age of 24, when 'the most blessed Lord had compassion on me and called me to his grace and sent my divine twin down to me ... He revealed to me the mystery of light and of darkness, the mystery of destruction ... the mystery of the creation of Adam, the first man. He also taught me the mystery of the tree of knowledge from which Adam ate and his eyes were opened.'

After these revelations, Mani was expelled from his Judeo-Christian monastery, outside Basra in Iraq, where he had lived since the age of 4. He then travelled east, to India and China, before returning to Persia to teach and, ultimately to martyrdom. During the twenty-six days of his marytrdom, he was staked out and was so laden down with chains that

The Prophet Mani – chosen at 24, martyred for 26 days

he could not move, so was starved to death, releasing his soul back to the great spirit on 2 March 274 AD. After his death his followers claimed that 'Mani khai' (Mani lives), which would later be elided into the word 'Manichees'.

24 ANGULAS MAKE A FOREARM ...

Twenty-four angulas make one hasta, which is one of the universal measurement units of mankind – the length of a forearm measured out to the extended middle finger. The hasta is a unit of measurement devised by the Harappan (the most ancient of India's urban civilisations along the Indus) and akin to the cubit used in Sumeria (the most ancient urban culture of Iraq) and ancient Egypt.

It seems that the basic Harappan unit was formed from the width of eight barley grains placed side by side, which was found to be equal to a finger's width (roughly 1.76cm). Ten of these finger-widths/barley-rows made an angula, while a dhanus (the length of a bow) was assessed as 108 of these finger-width/barley-rows. Anything with '108' in it was deemed to be very propitious in India and the East and so it was a favourite unit in which to design a citadel or a wall.

The use of barley as the ultimate foundation stone of measurement appears to be another universal element (alongside the forearm, the foot and the breadth of a finger), so that, for instance, you will find it underwriting the system of measurements used by the Vikings. But there has always been room for financial manipulation and speculation, especially from the great rival of barley, the slightly lighter wheat seed. Four wheat seeds equal three of barley, which are themselves considered to be on par with the seed from a carob tree.

... AND 24 PALMS MAKE A MAN

Four fingers make a palm, and six palms make a cubit, and four cubits make a man who should therefore be twenty-four palms in height. The other rule of male proportion is that, like the Emperor Charlemagne and King Edward I of England, we should stand six times the length of our foot. Half the length of the foot is also the extent of the average erect penis – which comes in at an average of just under six inches. A much greater mystery is whether the navel or the base of the penis is the centre of a man.

24 LETTERS OF THE GREEK ALPHABET

Alpha ❊ Beta ❊ Gamma ❊ Delta ❊ Epsilon ❊ Zeta
❊ Eta ❊ Theta ❊ Iota ❊ Kappa ❊ Lambda ❊ Mu ❊ Nu
❊ Xi ❊ Omicron ❊ Pi ❊ Rho ❊ Sigma ❊ Tau
❊ Upsilon ❊ Phi ❊ Chi ❊ Psi ❊ Omega

Through the medium of the Phoenician trading cities on the coast of Lebanon – places such as Tyre, Sidon, Byblos and Beirut – the ancient Greeks acquired the phonetic alphabet, which was passed on by them to all the other Western nations. Although the Greek letters synchronise with the twenty-four hours of the day and a pleasing correlation to twice the months of the year, the signs of the zodiac and the pantheon of gods, this numerical structure would not be retained. What is often forgotten is the range of size in the many derivative alphabets: for instance, Italian has 21 letters, Hebrew 22, English and French have 26, Dutch 27, Spanish 29 and Russian 33. This allows for an enormous flexibility in the results to be extracted from the pseudo-science of gematria (the counting up of numerical values from the letters of alphabet).

$$A\,\alpha \quad B\,\beta \quad \Gamma\,\gamma \quad \Delta\,\delta$$
$$E\,\epsilon \quad Z\,\zeta \quad H\,\eta \quad \Theta\,\theta$$
$$I\,\iota \quad K\,\kappa \quad \Lambda\,\lambda \quad M\,\mu$$
$$N\,\nu \quad \Xi\,\xi \quad O\,o \quad \Pi\,\pi$$
$$P\,\rho \quad \Sigma\,\sigma\varsigma \quad T\,\tau \quad Y\,\upsilon$$
$$\Phi\,\varphi \quad X\,\chi \quad \Psi\,\psi \quad \Omega\,\omega$$

23

THE 23 ENIGMA

In Tangier in 1960 the Beat writer William Burroughs met a sea captain called Captain Clark, who boasted to him that he had never had an accident in twenty-three years; later that day Clark's boat sank, killing him and everyone on board. Burroughs was reflecting on this, that same evening, when he heard a radio report about a plane crash in Florida: the pilot was another Captain Clark and the plane was Flight 23. From then on Burroughs began noting down incidents of the number 23, and wrote a short story, *23 Skidoo*.

Burroughs' friends Robert Anton Wilson and Robert Shea adopted the '23 Enigma' as a guiding principle in their conspiratorial *Illuminatus! Trilogy*. Twenty-threes come thick and fast: babies get 23 chromosomes from each parent; 23 in the I-Ching means 'breaking apart'; 23 is the psalm of choice at funerals; and so on. All nice examples of selective perception or, as Wilson put it, 'When you start looking for something you tend to find it.' The composer Alban Berg was also obsessed with the number, which appears repeatedly in his opera *Lulu* and in his violin concertos.

21

21-GUN SALUTE

One of the prime expressions of acknowledged sovereign national power is the twenty-one-gun salute, which seems to shows interesting analogies with the traditional coming of age of a fully entitled adult, who can vote, drink, serve in the army, have sex, marry and drive. But this age of adult initiation is only a very recent tradition in the Western world, coinciding with the end of university education, and is in any case today slipping back towards 18 and 16.

In fact, the twenty-one-gun salute has no spiritual origins. It evolved out of an expression of explosive power by the British navy that would demand a first salute from a foreign ship, then give them a withering demonstration of their superior discipline and power with their own salvo. Initially restricted to seven rounds, or seven cannon, it grew expediently with the size and arsenal of the ships of the line, but was capped at twenty-one so as not to waste too much time and powder. It also became less aggressive and by the nineteenth century ships would salute each other with a friendly gun-for-gun exchange.

20

20 FINGERS AND TOES

Twenty is perhaps the oldest, most natural large number for mankind to relate to, for it is the number we achieve by counting up all our fingers and toes. Echoes of this unit (called Vigesimal) can still be found in both the French and English language. The French still express eighty as 'quatre-vingts' (four twenties), while English keeps a special word ('score') for this number, as in the expression 'four score and ten'. And until decimalisation was introduced in 1971 the English monetary unit was still so ordered, with twenty shillings to the pound.

TOLKIEN'S 20 RINGS OF POWER

J.R.R. Tolkien's works are deeply embedded within a life-time of mythological and philological scholarship that merges strains of Celtic, Norse, Zoroastrian, Chinese and Byzantine storylines with his own imagination. At the heart of his *Lord of the Rings* trilogy is the Dark Lord Sauron, who

One Queen to rule them all. Twenty East African shillings from the 1950s.

has made twenty rings of power: Three for the Elves; Seven for the Dwarfs; Nine for the Kings of Men; and One, forged in Mount Doom, which will allow him to control all the nineteen ring wearers as explained by the secret rune verse, 'One Ring to rule them all, One Ring to find them, One Ring to bring them all, And in the darkness bind them.'

The 'Kings of Men' become the nine (another significant Tolkien number) dark riders – a mounted hit squad devoted to the service of the Dark Lord Sauron. Originally led by the witch-king of Angmar and the easterner Khamu, they were given rings to bind them into obedience to Sauron, and their character, shape and substance are gradually subsumed until they become spectral Nazgûl, 'ring-wraiths'.

19

SACRED 19 OF THE BAHAI

The much-persecuted Bahai community emerged in nineteenth-century Persia, though is today a diaspora of around 5 or 6 million spread around the world. They believe in the spiritual unity of mankind and revere Siyyid Ali Muhammad, who in 1844 foretold the appearance of a great prophet – who they believe was revealed in 1860 as Baha'u'llah ('Glory of God'). It was Baha'u'llah who compiled their Holy Book, which continues the spiritual development begun by Moses, Jesus, Muhammad and the other great prophets of the world.

The community has a special affinity to the number 19. They have a nineteen-day month, a nineteen-month year, an annual nineteen-day fast, and a communal festival every nineteen days with prayers and consultations. This is seemingly based on the numerical value of 'Wahid' – The One – though older associations exist, for 19 is the sum of the twelve signs of the zodiac together with the seven planets, reinforced by the nineteen-year-long Metonic cycle before the solar and lunar calendars exactly repeat themselves.

18

JEWISH LUCKY 18

Eighteen has long been a lucky number for Jews, for it is the numerical value of 'David' and, more important, for 'living'. So, if attending a wedding, a birthday or a Bar Mitzvah, you can achieve additional credit by writing a letter of congratulation in eighteen lines of text, evoking eighteen blessings or giving money in units of eighteen.

SACRED 18 OF THE WHIRLING DERVISH

The thirteenth-century Sufi mystic Rumi composed eighteen verses for the introduction of his iconic teaching verse, the Mathnavi, which like all his work must be referenced back to the Koran. For the single great introductory phrase to every Muslim verse, prayer and blessing is '*Bismi'llahi'r-rahmani'r-rahim*' – 'In the Name of God, Ever-merciful and All-forgiving' – which has eighteen consonants in it. A spiritual apprentice who wished to join the Mevlevi Sufi brotherhood (the Whirling Dervishes) was expected to first

Whirling Dervishes in Konya, 'The City of Rumi', around 1900.

learn to achieve eighteen kinds of service in the kitchen, each occupation requiring at least eighteen days of study. Similarly, the last ladder in the apprenticeship of learning was to meditate alone for eighteen days, having been escorted into one's cell by an eighteen-armed candelebrum. Gifts and courses of food were customarily served within the *tekke* (Dervish monastery) in sets of nine or eighteen.

17

HIDDEN 17 OF THE BEKTASHI

Just as the Mevlevi order of Dervishes could be identified with their fondness for the number 18, the Bektashi Dervishes were associated with 17. There were seventeen Bektashi saints who stood as the patrons of seventeen of the great trade guilds of the Ottoman Empire, while Ali ibn Abi Talib, their most important teacher, had seventeen companion disciples and offered seventeen sets of prayers three times a day.

The Bektashi smoothed over many of the differences between the Sunni and Shia schism within their own closely guarded belief structures, as well as delighting in numerology and absorbing various Turkic and Anatolian spiritual traditions. One of the oddest of these was the seventeenfold sacrifice to the guardian-spirit of Mount Ararat, where Noah's Ark was believed to have anchored. The flood was traditionally believed to have started on the seventeenth day of the second month and ended on the seventeenth day of the seventh month.

16

16 PROPHETIC DREAMS OF QUEEN TRISHALA

A white elephant ✳ A white bull ✳ A white lion ✳ Lakshmi ✳ Mandara flowers ✳ Silver moonbeams ✳ The radiant sun ✳ A jumping fish ✳ A golden pitcher ✳ A lake filled with lotus flowers ✳ An ocean of milk ✳ A celestial palace ✳ A vase as high as Mount Meru filled with gems ✳ A fire fed by sacrificial butter ✳ A ruby and diamond throne ✳ A celestial king ruling on earth

Queen Trishala is the mother of Mahavira, the last of the revered prophet teachers of the Jain. She was one of a family of seven princesses and one of her sisters became a Jain nun, while the other five married kings. Her husband was King Siddhartha, to whom she explained an extraordinary series of sixteen powerful dreams (fourteen in some accounts). Advised by seers, the King was able to tell her she was about to give birth to a strong, courageous son full of virtue.

15

15 RANKS OF THE KNIGHTS TEMPLAR

Grand Master ❊ Seneschal ❊ Commander of the Kingdom
of Jerusalem ❊ Commander of the City of Jerusalem
❊ Commander of Tripoli and Antioch ❊ Drapier ❊
Commander of Houses ❊ Commander of Knights ❊ Knight
Brothers ❊ Turcopolier ❊ Under Marshal ❊ Standard Bearer
❊ Sergeant Brothers ❊ Turcopoles ❊ Elderly Brothers

The Knights Templar were a crack force of armed monks,
established in 1129 to protect pilgrims journeying to
Jerusalem, and then employed to defend the Crusader
kingdoms of Outremer. After the fall of Outremer to
Turkic and Egyptian forces, the Templars no longer had a
function for a medieval Europe without any appetite for
crusading, and in 1312 they were suppressed by the Pope,
under pressure from the French King Philip IV. His reason
was straightforward: the throne was bankrupt and he wanted
the Order's considerable wealth – lands bequeathed to them,
priories in all the nations of Christendom and a banking
business. Because of the violence and suddenness of their
suppression (and the accusations of heresy levied against

them) a conspiratorial glamour continued to attach to the name of the Order, in contrast to its rival Hospitaller Knights of Saint John (who had the good sense to take over the island bases of Malta and Rhodes and still to an extent survive as a charitable institution). Indeed, the traditions of the Templars – or, to give them their full name, 'The Poor Fellow-Soldiers of Christ and the Temple of Solomon' – would be enthusiastically mined some 400 years later by the quasi-Templar Freemasonry Lodges established in Europe and North America.

During their heyday, the Templars' *Grand Master* was the absolute ruler over the Order and answered only to the Papacy. The *Seneschal* acted as both deputy and advisor to the Grand Master. The *Commander of the Kingdom of Jerusalem*, the *Commander of the City of Jerusalem* and the *Commander of Tripoli and Antioch* had the same powers as Grand Master within their own jurisdictions. The *Drapier* was in charge of the Templar garments. The *Commander of Houses* and the *Commander of Knights* acted as lieutenants to higher authorities within the Order. The *Knight Brothers* were the warrior-monks who wore the white tunic and red cross. Each was equipped with three horses and apprentice-like

squires. The *Turcopolier* commanded the brother sergeants in battle. The *Under Marshal* was in charge of the footmen and the equipment. The *Standard Bearer* was one of the sergeants and charged with carrying the order's banner. The *Sergeant Brothers* were warriors who did not have proof of eight quarterings of noble blood and thus had but one horse and no squires to assist them. The *Turcopoles* were local troops who would fight alongside the Templars. Sick and *Elderly Brothers* were no longer fit for active service but still members of the Order.

15 MEN ON THE DEAD MAN'S CHEST

Captain Flint of the Walrus ✻ Long John Silver ✻
'Blind' David Pew ✻ Billy Bones ✻ Ben Gunn ✻
Tom Morgan ✻ Job Anderson ✻ George Merry ✻
Dick Johnson ✻ Israel Hands

The nine men above are the named pirates alluded to in the song created by Robert Louis Stevenson in *Treasure Island*. Add the six sailors who were murdered by Captain Kiddand you get the full complement. And inside the dead man's chest? A cool £700,000 in gold.

'Fifteen men on the dead man's chest
Yo-ho-ho, and a bottle of rum!
Drink and the devil had done for the rest
…Yo-ho-ho, and a bottle of rum!'

14

GUARANTEED DEATH – AVOID 14

Fourteen is a number to avoid in any context in China and most of the Far East, for its tones sound like 'guaranteed death'. So do not bother looking for a 14th floor in an apartment block, number 14 in a row of houses, or the use of '14' in a number plate or telephone number. Other Chinese numbers to avoid, to a lesser extent, include 4 (which sounds like 'death'), 5 (which sounds like 'not'), and 6 (which sounds like 'decline'). And, as if to bear this out, in our world lives and teaches the fourteenth Dalai Lama, a spiritual hero fated to witness the slow death of his Tibetan homeland.

14 AND BACH

By giving each letter a number from its order in the alphabet you can deconstruct the name 'Bach' as follows: 2 for the B, 1 for the A, 3 for the C, 8 for the H – which makes 14. A pleasing mirror, or reversal, of this number can also be formed from 'J.S. Bach' – which gives 41. This pseudo-science of substituting numbers for letters is known as gematria (or *abjad* in Arabic) and has innumerable variations depending on whether you include vowels or which language you translate back to or transcribe into. It has often appealed to creative minds and may have been behind Bach's playful manipulation of the number 14, achieved by itself (in the fourteen canons of the Goldberg variations for instance) or in pairs of sevens that occur throughout his work.

Gematria is a very ancient tradition, particularly in the Near East, where it has often had official sanction, with poetic inscriptions commissioned by rulers to reveal the date of the publication of a book or the construction of a building. There are examples dating back to Sargon II of Assyria (in the eighth century BC). In the first century AD gematria became a recognised tool of Jewish hermeneutical scholarship and it was a tradition respected by many of the Ottoman Sultans. It seems only to have taken root in the imagination of Western Europe, however, in the seventeenth century.

14 STATIONS OF THE CROSS

Christ condemned to death ❈ Christ carries the cross ❈ Christ falls the first time ❈ Christ meets his mother ❈ Simon of Cyrene helps Christ carry his cross ❈ Veronica wipes the face of Christ ❈ Christ falls a second time ❈ Christ meets the women of Jerusalem ❈ Christ falls a third time ❈ Christ's clothes are removed ❈ Christ is nailed to the cross ❈ Christ dies on the cross ❈ Christ's body is taken down from the cross ❈ Christ is laid in the tomb

These fourteen Stations of the Cross may be familiar from the little altar-like plaques or images attached to the naves of all modern Roman Catholic churches and the more ritualistically inclined Anglican and Lutheran churches.

They grew out of the habit of creating pilgrimage-like replicas of the Via Crucis/Via Dolorosa that a pilgrim to Jerusalem would trace on their spiritual journey to the Place of Crucifixion. The practice goes back to the fifth century, in part to offer a substitute for the pilgrimage in times of trouble, but also to offer it up as part of the Lenten cycle of events leading up to the central Christian festival of Easter. It was popularised by the Franciscans until, by the eighteenth century, it had become an essential aspect of worship, when the number of stations, which could fluctuate between six and thirty, was fixed at fourteen.

13

13 AT A TABLE – AND THE 13TH MONTH

Thirteen is a famously unlucky number in the Western world. I certainly grew up with the belief that to invite thirteen guests to sit around the table doomed the last to some nameless dread – so, to avoid that fate, our table was always laid to include fourteen. It was a belief shared by Napoleon, F.D. Roosevelt and John Paul Getty, and concern over the number 13 is the most common form of Western superstition. Hotels often have no room 13, tower blocks tend to avoid a 13th floor, and travel agents know that the thirteenth of the month (especially if it falls on a Friday) will be short of bookings.

The most common explanation for unlucky thirteen is the Last Supper, where thirteen sat down to eat, one of whom was a traitor plotting the arrest and judicial murder of his host and master. But similar stories can be found in many other cultures, such as the Viking Norse, who remembered how Loki stumbled into a gathering of twelve gods (from which he had been excluded) and in his envy started plotting the events that would lead to the end of the world.

Robert Graves enthusiastically listed in *The White Goddess* the various mythological companies of thirteen that tend to lead to the betrayal, if not sacrificial death, of one of their members: be they Arthur and his twelve knights, Odysseus and his twelve companions, Romulus and the twelve shepherds, Roland and the twelve peers of France, Jacob and his twelve sons, or Danish Hrolf and his twelve Berserks. Not to mention the thirteen dismembered portions of Osiris's body recovered by Isis from the Nile.

The ultimate cause of our attitude to thirteen may be that the thirteenth month of the year was always weak and withered. For, although twelve lunar months almost fill up our solar year (to produce 360 days from twelve sets of 29 and a half days) there was always the issue of a left-over period of five days. This was considered in ancient cultures to be the thirteenth month, a five-day oddity, often believed to be a period of immensely bad luck where the world was not policed by the normal powers, and evil spirits held brief reign. Some cultures made this into a Saturnalia-like carnival, where the normal roles of society were reversed; others deemed it a needful time for sacrifice.

13 INSTRUMENTS OF THE PASSION

Torch of Malchus (and his ear cut off by Peter) ❈
Pilate's Jug and Bowl ❈ Pillar on which Christ was bound ❈
Whip ❈ Crown of Thorns ❈ Cross ❈ Nails ❈ Sponge of Gall
❈ Lance ❈ Purse around Judas's neck ❈ Cock that crowed
thrice ❈ Ladders and pincers at the deposition ❈
Dice cast by the soldiers

These were popular from the thirteenth century as heraldic emblems of the Passion, though their number can grow with the addition of the Holy Napkin, Saint Peter's sword, a Blindfold, a hand striking and a head spitting.

SATAN'S 13 PEERS OF HELL

Beelzebub ❋ Moloch ❋ Chemos ❋ Peor ❋ Baalem ❋
Ashtoreth/Astarte ❋ Thammuz/Adonis ❋ Dagon ❋
Rimmon ❋ Osiris ❋ Isis ❋ Horus ❋ Belial

In Book One of Milton's *Paradise Lost*, Satan, having been expelled from Heaven, falls 'nine times the space that measures day and night' into Hell's cavern. In this 'dismal situation waste and wild, a dungeon horrible on all sides round as one great furnace flamed' he rears up from a pool of liquid fire to offer words of comfort to the fallen cherubs. One by one, Milton identifies and to a certain extent creates the thirteen chief captains of Hell, from his own selective reading of the mythology of the ancient Near East, who follow 'their great Emperor's call' in order to stand beside him. These Peers of Hell are a bad lot – 'besmeared with blood', fomentors of 'lustful orgies' and 'wanton passions in the sacred porch' – and summon myriad other fallen angels to arms with a shout that 'frighted the reign of Chaos and old Night'.

13 HALLOWS OF BRITAIN

Dyrnwyn (sword of Rhydderch Hael) ❋ Basket of
Gwyddno Garanhir ❋ Horn of Bran Galed ❋ Platter of
Rhegynydd Ysgolhaig ❋ Chariot of Morgan Mwynawr ❋
Halter of Clydno Eiddyn ❋ Knife of Llawfrodedd Farchawg
❋ Cauldron of Tyrnog ❋ Whetstone of Tudwal Tudglyd ❋
Robe of Padarn Beisrudd ❋ Mantle of Tegau Eururon ❋
Chessboard of Gwenddoleu ❋ Mantle of Arthur

This list from the medieval *Mabinogin* manuscript gives us a precious insight into the heroic Iron Age culture of ancient Britain as it survived in Wales. Though found in a collection of thirteenth-century manuscripts bound into two books,

The Dyrnwyn legend interpreted Japanese anime style

the eleven stories clearly predate Geoffrey of Monmouth's skilful retelling of the legend of Arthur, and the *Mabinogin* was also a source that provided Tolkien with rich imagery. It cast shades, too, over J.K. Rowling's *Harry Potter*, and has inspired Japanese anime.

Among its legends, the most potent, perhaps, were those of *Dyrnwyn* – the 'white hilt' sword of Rhydderch Hael, which, if any man except Hael drew it, would burst into a flame from point to hilt; and the *mantle of Arthur*, which was basically an invisibility cloak. Other magical clothing came in the form of the *Robe of Padarn Beisrudd*, which fitted every one of gentle birth (but no churl could wear it), and the *Mantle of Tegau Eururon*, which only fitted ladies whose conduct was irreproachable.

Many of the stories were about the provision of plenty: if food for one man was put into the *Basket of Gwyddno*

Garanhir, it would suffice for a hundred; the *Horn of Bran Galed* always found the very beverage that each drinker most desired; the *Platter of Rhegynydd Ysgolhaig* always contained the very food that the eater most liked; the *Knife of Llawfrodedd Farchawg* would serve twenty-four men simultaneously at any meal; while the *Cauldron of Tyrnog* would instantly cook any meat put in for a brave man, but never boil for a coward. In similar vein, if the sword of a brave man was sharpened on the *Whetstone of Tudwal Tudglyd* its cut was certain death, but if of a coward, the cut was harmless.

Futher wish-fulfilment arrived in the form of the *Chariot of Morgan Mwynvawr*, which transported him in a moment, wherever he desired, and the *Halter of Clydno Eiddyn*, which provided whatever horse he wanted.

Finally – and most Potter-like – there was the *Chessboard of Gwendolleu* –a golden chessboard with men of silver. When the men were placed upon it, they would play themselves.

13 BARS ON THE UNION FLAG

Delaware ❊ Pennsylvania ❊ New Jersey ❊ Georgia ❊ Connecticut ❊ Massachusetts ❊ Maryland ❊ South Carolina ❊ New Hampshire ❊ Virginia ❊ New York ❊ North Carolina ❊ Rhode Island

These thirteen states rebelled against the British Crown in the eighteenth century and are represented by the thirteen parallel bars of the Union Flag. Curiously the Founding Fathers adapted their design from a flag used by the East India Company, which had created its own version of the mercantile Red Ensign with thirteen red and white parallel bars instead of a plain red field. On an early version, thirteen stars complemented the bars, then more and more stars were added as the American state expanded into the Indian lands to their west.

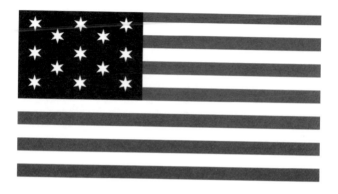

On 14 June 1777, the Second Continental Congress passed the Flag Resolution, which stated: 'Resolved, That the flag of the thirteen United States be thirteen stripes, alternate red and white; that the union be thirteen stars, white in a blue field, representing a new constellation.'

Another curiosity is that, just as thirteen states rebelled against their royal motherland, it would be thirteen states that would later rebel against the republican union:

South Carolina ❋ Mississippi ❋ Florida ❋ Alabama ❋ Georgia ❋ Louisiana ❋ Texas ❋ Virginia ❋ Arkansas ❋ Tennessee ❋ North Carolina ❋ Missouri ❋ Kentucky

It is not always remembered that all thirteen of the Confederate states seceded from the Union between December 1860 and December 1861, after the process of a democratic vote, either a popular referendum or a vote from their House of Representatives. They had imagined the Union was a free compact, which they were entitled to leave, just as they had joined it. After four years of war, the northern armies of the Union had either destroyed or occupied all the territories of those who had attempted to secede.

12

POWER OF 12

One of the cornerstones of human life is that there are twelve months in a year. Recent archaeological discoveries suggest that we have been notching off the days of the cycle of the moon for hundreds of thousands of years, using stone tools to mark bone. And it must have been one of our first pieces of inherited science that the counting off of twelve moons fitted magically into the annual miracle of the changing seasons. As there are (very nearly) thirty days in each lunar month, one of the very first joys of multiplication must have been that when multiplying these twelve months by thirty, you create 360, which is (roughly) how many days there are in the year. So we have always divided up the heavens – and any circles we come across – into 360 degrees.

The added harmony of the tides, and the female cycle of fertility fitting into the lunar months, provided further proof that there was a pattern and an order to the world. And one of those patterns was very clearly that twelve moons make one year. This innate power of twelve was further reinforced when the heavens, through which the sun was imagined to process, were also neatly divided into twelve segments. Each of the twelve signs of the Zodiac were allotted 30 degrees of the Heavenly circle very early on in mankind's construction of an ordered world. This would later be reinforced by other twelvefold divisions, aspiring to create the same graceful, ordered inevitability.

These twelvefold divisions of the night sky and the moon also made for very easy organisation. A clan or a district could became associated with a particular month, and so, whether it was taking turns to guard a citadel, provide food for a shrine or furnish a choir for the temple at the next full moon, it became almost a natural habit of mankind to form themselves into twelve.

12 SIGNS OF THE ZODIAC

Aries (Ram) ✳ Taurus (Bull) ✳ Gemini (Twins) ✳ Cancer (Crab) ✳ Leo (Lion) ✳ Virgo (Virgin) ✳ Libra (Scales) ✳ Scorpio (Scorpion) ✳ Sagittarius (Archer) ✳ Capricorn (Goat) ✳ Aquarius (Water Carrier) ✳ Pisces (Fish)

The zodiac is a very old concept, which has impregnated our thought patterns for thousands of years. In essence it was the observation of the sun's circular path through the heavens (as viewed from the earth) and the division of this into twelve equal sections of 30 degrees to make a complete circuit of 360 degrees. Like so much of our world, the start date is spring, the vernal equinox of 21 March, so Aries (21 March–20 April) must always start the cycle.

The symbols chosen by the Sumerian astrologers and their imaginative pattern-making of sacred shapes from the most prominent stars passed seamlessly into Babylonian, Egyptian, Hindu and Greek thought – notably through the teachings of a pair of well-travelled Greeks, Eudoxus of Cnidus and

*The Millenial Order of the Twelve Signs of the Zodiac
in a modern neo-pagan dressing.*

from the Egyptian-Greek scholar Ptolemy, whose Almagest colonised the imagination of both Islam and Christendom.

But just to read the Sumerian names is to stand in witness of an impressive piece of 5,000-year-old living continuity: Luhunga (Farmer) is Aries; Gu Anna (Bull of Heaven) is Taurus; Mastabba Bagal (Great Twins) is Gemini; Al-Lul (Crayfish) is Cancer; Urgula (Lion) is Leo; Ab Sin (virgin land) is Virgo; Zib Baanna (scales) is Libra; Girtab (scorpion) is Scorpio; Pabilsag (soldier) is Sagittarius; Suhurmas (goat-fish) is Capricorn; Gu La ('Great One') is Aquarius, the water bearer during the winter rains; and Dununu (fish cord) is Pisces.

12 MONTHS OF THE ATTIC CALENDAR

Elaphebolion (March/April) ⁕ Mounichion (April/May)
⁕ Thargelion (May/June) ⁕ Skirophorion (June/July) ⁕
Hekatombaion (July/Aug) ⁕ Metageitnion (Aug/Sept)
⁕ Boedromion (Sept/Oct) ⁕ Pyanepsion (Oct/Nov)
⁕ Maimakterion (Nov/Dec) ⁕ Poseideon (Dec/Jan) ⁕
Gamelion (Jan/Feb) ⁕ Anthesterion(Feb/March)

One of the wonders of ancient Greece was that, despite a shared civilisation that was held in common (and so influenced Western civilisations), each region, sometimes each city, also kept stubbornly to its own traditions – even its own calendar. For instance, in moving between Delphi, Thebes, Athens, Crete, Macedonia, Epirus, Sicily or Sparta, you would move between different sacred vocabularies, not to mention different dating systems based on the foundation date of a city. However, all of them leant on the ancient Babylonian division of the four seasons and the twelve lunar months, so there was also room for easy understanding.

12 ROMAN MONTHS OF OUR YEAR

January ⁕ February ⁕ March ⁕ April ⁕
May ⁕ June ⁕ July ⁕ August ⁕
September ⁕ October ⁕ November ⁕ December

Our English months are a straight textual survival from the pagan Roman Empire due to the literate record-keepers of the Church staying true to the Julian calendar. Nothing Celtic, Anglo-Saxon, Norse, let alone Christian or Norman-French, can be identified from their names.

The Romans initially only named ten of their months and most of them were given pragmatic but unpoetic numbers. They also originally left two of them unnamed, seemingly because they were so unpropitious at the end of the year that they were better not written down. But King Numa bit the bullet of calendar reform in 713 BC and added the names of January and February, as well as the thirteenth odd month of five days, the unlucky Mercedonus. This was gradually merged into February, though in a very unsystematic way, until in 46 BC Julius Caesar listened to the scholars and tidied things up so that the months once more kept pace with the seasons. With a further bit of housekeeping by Pope Gregory in 1583 (who nipped off eleven days for the same reason), this calendar has served us to this day.

January was named after Janus, the Roman double-faced god of entrances. *February* was the Roman month of purification, which we believe was celebrated on the 15th. *March* was for a long time the beginning of the year (indeed for tax matters it still is) and named after Mars, god of war. *April* is from Aprilis, a shortened version of Aphrodite, but not curiously from Venus, the Roman title for the goddess of love and fertility. *May* is from the Latin 'Maius', Maia's month – one of the Roman titles of the fecund daughter of the Great Mother Goddess who we might know better as Persephone, the spirit of spring. *June* is from Latin 'Iunius', Juno's month, the Great Mother Goddess herself, elsewhere addressed as Hera or Ceres.

July was originally known as Quintilis (the fifth month) but would later be named after Julius Caesar. *August* was once known as Sextilis, the six month but renamed Augustus in 8 BC after the Emperor Octavian, great-nephew of Julius Caesar. *September* was originally the seventh month, but Tiberius, much to his credit, stopped the sycophantic rot by refusing to have it named after himself. *October* is yet another Latin misnomer, for their pragmatically named eighth month now serves as our tenth. *November* was their ninth month, now doing service as our eleventh. *December* was their tenth month and is now our twelfth.

12 DAYS OF CHRISTMAS

Twelve Drummers Drumming ❋ Eleven Pipers Piping ❋
Ten Lord's a-Leaping ❋ Nine Ladies Dancing ❋
Eight Maids a-Milking ❋ Seven Swans a-Swimming ❋
Six Geese a-Laying ❋ Five Gold Rings ❋ Four Calling
Birds ❋ Three French Hens ❋ Two Turtle Doves ❋
And a Partridge in a Pear Tree

This ancient Christmas song is full of mysterious allusions that seem to mix the sacred with the profane. Its elements – like those in the similarly-aged song 'Green Grow the Rushes O' – have been piously interpreted as: the twelve Apostles, the eleven faithful disciples at the Last Supper, the Ten Commandments, the nine fruits of the spirit, the eight beatitudes, the seven gifts of the Holy Ghost, the six days of Creation, the five Books of Moses, the four Gospels, the three virtues of faith, hope and charity, the two holy books of the Old and New Testaments, and the One True God.

The related song 'Green Grow the Rushes O' appears more explicitly Christian, but, once again, the symbolism behind Nine, Eight, Seven, Six and Five still seems a little forced.

I'll sing you twelve, O / Green grow the rushes, O / What are your twelve, O? / Twelve for the twelve Apostles / Eleven for the eleven who went to Heaven / Ten for the

ten commandments / Nine for the nine bright shiners /
Eight for the eight bold rangers / Seven for the seven stars
in the sky / Six for the six proud walkers / Five for the
symbols at your door / Four for the Gospel makers /
Three, Three, the rivals / Two, two, lily-white boys /
Clothed all in green, O / One is one and all alone /
And evermore shall be so.

12 DAYS OF THE NOWRUZ FESTIVAL

This New Year festival has Zoroastrian roots and is associated principally with Iran, but it is celebrated from Syria to India and across all of Central Asia, the Caucasus, Kurdistan and Turkey. Its rituals vary widely but most are based around a twelve-day succession of events. This can begin with great bonfires fed all night to symbolize the victory of light over winter darkness, then the spring cleaning of the house, the bringing into the house of something green (like a palm tree or a fir tree – depending on latitude) around which a vigil of candles may be lit, then the making of a splendid feast full of special seasonal foods, including displays of dried fruits and nuts, an exchange of gifts between close family members, followed by an exchange of visits between neighbours and cousins. In some regions there followed a traditional 'period of misrule', where men would dress as women, and women as men, children would lord it over adults and the poor would be served by the rich and the powerful would be publicly mocked by licensed fools. On the thirteenth day, the festival concludes with a family picnic, with music and dance and the quiet contemplation of the beauties of nature and some thought for future marriages and the exchange of such symbols of fertility as coloured eggs.

Many of these Zoroastrian practices were mirrored in the festival of the winter solstice or the Roman-era cult of the

This bas-relief from Persepolis, in Iran, is thought to show a Zoroastrian Nowruz festival procession.

unconquered sun. They would get absorbed wholesale into the Christian Easter and Christmas festivals, for Christ's birthday was fused with the winter solstice, just as his death was tied to the spring festival.

12 DISCIPLES AT THE LAST SUPPER

Simon-Peter ✳ Andrew ✳ James ✳ John ✳ Thomas ✳
James ✳ Philip ✳ Bartholomew ✳ Matthew ✳
Simon the Zealot ✳ Thaddaeus (Jude) ✳ Judas Iscariot

The Twelve Disciples of the Last Supper would, with two changes (exit Judas and enter Matthias – who was chosen by lot), become the twelve Apostles.

The Apostles, from the Greek for 'messenger', were the disciples who became missionaries after the crucifixion and resurrection of Jesus. They were at first symbolised as sheep in church imagery, but from the sixth century as individual men, each with their own particular symbols: Simon, renamed *Peter* (keys, fish, upside down cross); *Andrew* his brother (saltire cross); *James the Greater*, the son of Zebedee and brother of Saint John (pilgrim's staff and/or sword); *John* the beloved disciple (eagle or chalice with snake); *Philip* (cross/cross staff); *Bartholomew* (skinner's knife); *Thomas-Didymus* 'Doubting Thomas' (set square); *James the Lesser*, son of Alphaeus (fuller's club); *Matthew* the tax collector (purse); *Simon the Zealot* (saw); *Thaddaeus* also known as *Jude* (image of Jesus and flame); and *Matthias* (lance).

There is substantial agreement between the Gospels about the names and the numbers of the 'twelve' who, with the exception of John, all paid for their abundant faith with their lives. But confusion can be caused as there is an addition cast of further Apostles, traditionally set at seventy or seventy-two, who include such characters from Acts and Epistles as Saints Paul, Barnabas, Mark and Titus, and which may have included known female followers of Jesus such as Mary Magdalene, who have been almost airbrushed out of established church history.

12 SIBYLS

Persian/Hebrew/Palestinian ❖ Egyptian/Baylonian ❖ Libyan ❖ Delphic ❖ Cimmerian ❖ Erythrean ❖ Samian/ Herophile ❖ Cumaean ❖ Hellespontine/Trojan ❖ Phrygian ❖ Etruscan/Apennine ❖ Tiburtine

The Sibyls, the female prophets of the pagan world, remained enduringly mysterious and popular figures in Christendom, largely because they were considered to have prophesied the end of the old pagan world and the coming of Christ.

The most famous of the Sibyls were those connected with the great oracular shrines of Apollo at Delphi and of Zeus Ammon in the Saharan oasis of Siwa. But the prophetess of Erythrae (in western Anatolia, on the shore opposite Hios) is credited with the invention of the acrostic, and the son of the Cimmerian Sibyl clearly had sufficient spiritual authority to establish the first shrine to Pan at Rome.

However, it is the Cumaean Sibyl who walks into the pages of history with the most confident stride. It was her collected prophecies in nine volumes, written in Oscan (one of the ancient languages of southern Italy), that were said to have been offered to King Lucius Tarquinius Superbus for a fabulous sum. When he indignantly refused, she returned three years later, looking younger, and asking the same price, despite having destroyed three of the volumes. When he once again refused, she returned a third time, having burnt a further three of the bound volumes, but this time looked even younger, and once again asked exactly the same price for the remaining three volumes. The King at last realised the importance of what he was being offered and hurried to acquire the precious texts at any price. They became known as the Sibylline Books and were stored in the Capitoline Temple with the other holies of holies, and could only be consulted by a college of fifteen priests acting together on the direct instructions of the Senate, until tragically they were destroyed by fire in 83 BC. The Romans later tried to reassemble these lost books of prophecies and wisdom by sending out ambassadors to record the utterances of all the other historical Sibyls who could be traced. This collection, composed of Greek hexameters (so probably more than a little like the I-Ching of ancient China) survived until it was destroyed by a Christian general in 405 AD.

Virgil may have been given access to some of this collection, for in his Fourth Eclogue he creates the prophecy of 'a new breed of men sent down from Heaven' (probably no more than a sycophantic reference to Augustus), which was later taken to predict the coming of Christ.

12 OLYMPIAN GODS

Zeus (Jupiter) ❊ Hera (Juno) ❊ Poseidon (Neptune) ❊ Aphrodite (Venus) ❊ Athena (Minerva) ❊ Apollo (Apollo) ❊ Artemis (Artemis) ❊ Hermes (Mercury) ❊ Dionysus (Bacchus) ❊ Hades (Pluto) ❊ Aries (Mars) ❊ Hephaestus (Vulcan) ❊ Hestia (Vesta)

The cult of six female goddesses paired with six male gods came to Greece from western Anatolia during the Iron Age, though the doubling up of a trinity of female goddesses, and then giving them appropriate male counterparts, was a familiar aspect of many ancient cultures. Hercules is first credited with organising sacrifices to all twelve of the great gods who dwelt on Mount Olympus and the oldest such altar was associated with Athens. The precise names of the Olympian pantheon would shift during 1,000 years of worship, which is why thirteen deities have been listed. Hades, as Lord of the Underworld, was vulnerable to downgrading, especially in favour of Hestia/Vesta, while at other times Hephaestus could be exchanged for Hercules, and in the early period Dionysus, held an equivocal position.

THE CITY'S 12 GREAT LIVERY COMPANIES

Mercers ✣ Grocers ✣ Drapers ✣ Fishmongers ✣
Goldsmiths ✣ Merchant Taylors ✣ Skinners ✣ Haberdashers
✣ Salters ✣ Ironmongers ✣ Vintners ✣ Clothworkers

Medieval London was a free city that governed itself through the interconnection between its wards, its parishes and the guilds that controlled the various different aspects of trade. The twelve great livery companies are the richest and oldest of the guilds whose foundation charters (though often much older) can be securely dated to fourteenth-century documents. They were (and are) managed by a clerk but controlled by a Master, a number of wardens and a court of assistants elected by the liverymen and freemen of the company. Access is through patrimony (descent), servitude (apprenticeship to a guild master) or redemption (a fee).

Liverymen famously squabbled about order of precedence. It is said the origin of the phrase 'being at sixes and sevens' is the Skinners' and Merchant Taylors' dispute and eventual agreement to exchange being number 6 and 7 in the hierarchy.

CHARLEMAGNE'S 12 PALADINS

Roland ✣ Oliver de Vienne ✣ Naimon of Bavaria
✣ Archbishop Turpin ✣ Ogier the Dane ✣ Huon de
Bordeaux ✣ Fierabras ✣ Renaud de Montauban ✣
Ganelon ✣ Guy de Bourgogne ✣ William of Gellone ✣
Girart de Roussillon ✣ Aymeri de Narbonne

The Twelve Paladins (or Twelve Peers) dominated the imagination of medieval Europe for at least 500 years. At the heart of the story is a band of twelve noble knights who assist the Emperor Charlemagne in defending Christendom from the assaults of Saracens from the south (especially Muslim

Spain) and pagans from the north. Into this central epic are woven fragments of Norse and classical mythology, doomed love, chivalric duels, legendary quests, as well as real battles transformed into romantic legend. This bundle of stories is known as *The Matter of France* and is consciously interlinked with *The Matter of Britain* (the Arthurian cycle of tales) and *The Matter of Rome*.

There are many variant lists of the Twelve Peers but the first seven given here have to be included. *Roland*, an historical marcher-lord of the Carolingian Breton frontier, fated to die protecting the Christian army at the battle fought at the Roncevaux Pass, is at the centre of the tale. Key tales recount how he won his horse Veillantif, his magical sword Durendal and his battle-horn, Oliphant. Second in chivalric glory is *Oliver*, brother of Roland's love, Aude. *Naimon* is the German straight guy, Charlemagne's most dependable soldier and father of Sir Bertram. *Archbishop Turpin of Reims* is a historical figure who died in 800 AD fused with another warrior cleric. *Ogier* is both a knight-errant and the once and future king of Denmark, asleep beneath Kronborg Castle wrapped up in his beard. *Huon* is set a series of near impossible quests by his emperor to cleanse him of the blood-guilt of killing Prince Charlot. *Fierabras* is the Saracen champion who converts to Christianity.

Renaud is another major figure, supported by three brothers (Alard, Guiscard and Richard), a magical sword (Froberge) and a magical horse (Bayard). *Ganelon* is the Judas-like traitor within the band of twelve brothers who will be torn apart by four wild horses. *Guy de Bourgogne* marries the Saracen beauty Floripas (sister of Fierabras). And *William of Gellone* is the archetypally adventurous second son who advances himself to become the Marquis Court Nez.

But unifying all of these characters is their purity and chivalry. So a Spanish soldier about to be executed on the banks of the Rio Plate in Argentina in 1536 could look his commander in the eye and declare, 'Some days things will as God wills, and the Twelve Peers will rule,' and know that these last words would be remembered by his comrades.

12 KNIGHTS OF THE ROUND TABLE

Lancelot du Lac ❖ Kay ❖ Galahad ❖ Perceval ❖
Tristram ❖ Gawain ❖ Gareth ❖ Lamorak ❖ Gaheris ❖
Mordred ❖ Bors ❖ Bedivere

It was the linkage of the legend cycle of Arthur with that of Charlemagne (above) that seems to have encouraged an early listing of twelve knights of the Round Table. But there are over 100 named knights associated with the Arthurian legends, and tables of thirty-four and fifty knights noted. Those listed above, however, are the Round Table's twelve chief characters. They

*King Arthur's Twelve Knights – and a few hangers-on – depicted at the
Round Table in a fifteenth-century engraving.*

include knights close or related to Arthur, such as his foster brother Sir Kay, his nephews Sir Gawain and Sir Gaheris, and his illegitimate son and nemesis, Sir Mordred.

12 ONCE AND FUTURE KINGS

Emperor Constantine XI ❋ Emperor Charlemagne ❋ Emperor Frederick Barbarossa ❋ King Arthur of Britain ❋ King Dom Sebastian of Portugal ❋ Fionn Mac Cumhail, High King of Ireland ❋ Ogier, King of Denmark ❋ King Wenceslas of Bohemia-Prague ❋ King Matthias Corvinus of Hungary ❋ Prince Csaba ❋ Vainamoinen of Finland ❋ King Lacplesis of Latvia

In Norse tradition, the Kings under the Mountains are the great warriors of the world who will join the king of the gods in the last battle. They parallel the Christian belief that utter chaos will engulf the land before Jesus returns in glory as the Messiah. In the wake of these traditions are numerous tales of 'once and future kings'. Many of them relate to the tales of the Paladins. They may require a legendary sword to be found and thrown back into a sacred river or lake as a symbol that the hour has come, or a time-worn flag to be unfurled or an ancient horn to be sounded. Others speak of boys trespassing into hidden caves who come across a great bearded king who asks solemnly, 'Do the ravens still circle the mountain?' and, being told that they do, mutters, 'My time is not yet nigh.'

The *Emperor Constantine XI* died defending the walls of Constantinople against the Ottoman Turks but his body was never discovered and is believed to be held in an enchanted cave awaiting the time for him to restore the Byzantine Empire. *Charlemagne* lies beneath the Untersberg awaiting Christendom's hour of greatest need. The body of *Frederick Barbarossa* was swept away at the end of his peaceful crusade but will come again. The boy king *Dom Sebastian of Portugal* died heroically at the end of the battle of the Three Kings (1578), which destroyed the Empire of Portugal at its zenith, and will rise once more in the mythology of Sebastianists. So too with *King Arthur*, whose Messianic return has fired Britons (and especially the Welsh and Cornish) since the twelfth century.

Fionn will rise again at Ireland's most testing hour, summoned when the Dorn Finn horn sounds three times. The bearded knight *King Ogier* lies asleep below the vaults of Kronborg Castle awaiting Denmark's call. *King Wenceslas* lies in the roots of Mount Blanik awaiting the time to ride out on his white horse wielding the magical sword of Bruncvik. *King Matthias* will return to lead Hungary to its days of glory, just as *Prince Csaba*, the son of Attila the Hun, lies in the mountains ready to assist the Szekelys marcher lords left behind by the Horde. *Vainamoinen of Finland* sails mysteriously out of his national epic (the *Kalevala*) to await the last summons. *Lacplesis* lies in his secret tomb beneath the banks of the River Daugava.

12 WOMEN OF THE PROPHET

Khadija ❀ Sawda ❀ Aisha ❀ Hafsa ❀ Zaynab bint Khuzayma ❀ Umm Salama ❀ Zaynab bint Jahsh ❀ Juwayriyya ❀ Safiya ❀ Rayhana ❀ Umm Habiba ❀ Maymuna ❀ Maria or Meriem

Khadija fell in love with the young Muhammad for many reasons: 'I love you because you are well centred, not being a partisan amongst the people for this or for that; you are trustworthy and have a beauty of character and I love the truth of your character.' She was an intelligent, wealthy widow who bore him all four of his surviving children and assisted him in the first testing years of his spiritual mission. He would take no other wife while she was alive, and he was affectionately taunted towards the end of his life (by one of his wives) for still wishing to be alone with Khadija in a simple reed hut.

Sawda was a widow of 30 years old who had been one of the first Muslims to escape the persecution from pagan Mecca; she was more a housekeeper than a companion. *Aisha* was Muhammad's only virgin bride. The beautiful headstrong daughter of his best friend, Abu Bakr, she never conceived a child of her own and became locked into an unprofitable rivalry with the Prophet's daughter Fatimah and his cousin-son-in-law Ali. *Hafsa* was the daughter of Omar, whose first husband died at the battle of the wells of Badr; she was fiery-tempered, intellectual and possessed the first written version of the Koran, before even that compiled by the Caliph Uthman.

Zaynab bint Khuzayma was the daughter of a Bedouin nomad, famously generous to the poor, but died after her first few months at the Prophet's house. *Umm Salama* was the widow of Muhammad's first cousin, Abu Salama, who had died at the battle of Uhud; she proved a valued counsellor and mediator. *Zaynab bint Jahsh* was a controversial partner, for she had been the wife of his adopted son, his ex-slaveboy Zayd, who had been a trusted military commander.

Juwayriyya was a daughter of a Bedouin chief. *Safiya* was the daughter of Sheikh Huyay, leader of a Jewish-Arab clan of Medina. *Rayhana* was a concubine wife of the Prophet who never coverted to Islam; she was presented to him after all the males of her Jewish clan had been executed. *Umm Habiba* was a diplomatic marriage, for she came from the Quraysh tribe who dominated pre-Islamic Mecca and led the pagan resistance to Muhammad. *Maymuna* was the widowed sister-in-law to Muhammad's clever banking uncle Abbas.

Finally, *Maria* or *Meriem* was one of two Coptic concubines brought to Medina as gifts from the governor of Egypt. She was given her freedom after she gave birth to Muhammad's son Ibrahim, yet she was never known by the honorific of 'Mother of the Faithful', perhaps because of her past as a dancing girl and Christian.

12 LABOURS OF HERCULES

Slay the Nemean Lion ❖ Kill the Hydra of Lake Lerna ❖ Capture the Ceryneian Hind ❖ Capture the Erymanthian Boar ❖ Cleanse the Augean Stables ❖ Kill the Stymphalian Birds ❖ Capture the Cretan Bull ❖ Kill the Man-Eating Mares of King Diomedes of Thessaly ❖ Seize the Girdle of Hippolyte ❖ Capture the Cattle of Geryon ❖ Take the Golden Apples of the Hesperides ❖ Capture Cerberus, the Three-Headed Guardian of the Underworld

Hercules was sent by the Delphic Oracle to labour for twelve years for the cowardly and unworthy King Eurystheus – the punishment for the blood-guilt of murdering his own six sons during a bout of divinely afflicted madness. The King initially set Hercules ten labours, but these were stretched by a further two (to make our dodecathlon) by Eurystheus's thankless quibbling. Each of the labours was achieved through an admirable mixture of brute strength and cunning. Many have been connected either with ancient cult places of the Great Goddess in the Peloponnese or with the

Hercules in action, with a Lion of Nemea cloak slung over his shoulder.

sort of places that his Phoenician role model Melkarth, the hero-king of Tyre, would have visited in the early Iron Age.

Beneath the fun and dash of the adventure stories there lurks some earlier Gilgamesh-like narrative of a doomed mortal hero struggling against his ultimate, sorry fate – which was to watch his wife being raped, accidentally kill her, then burn his own body with a poisoned cloak that drove him in an extremity of pain to climb onto a funeral pyre.

12 MONTHS OF THE FRENCH REPUBLICAN CALENDAR

Vendémiaire (grape harvest) ❊ Brumaire (fog) ❊
Frimaire (frost) ❊ Nivose (snowy) ❊ Pluviôse (rainy) ❊
Ventôse (windy) ❊ Germinal (germination) ❊
Floréal (flower) ❊ Prairial (pasture) ❊ Messidor (harvest) ❊
Thermidor (heat) ❊ Fructidor (fruit)

This calendar was part of a reform movement to make over the world into a rational but yet poetic place. Its first month, Vendémiaire (from the Latin for 'grape harvest') started the day after the autumn equinox, which was neat, for it was also the day after the abolition of the monarchy on Year 1 of the Republic, 22 September 1792.

The poet-journalist Fabre d'Eglantine was called in to advise the calendar committee on the naming of the months. They were to be exactly thirty days long, composed of three ten-day long weeks, each ending with a *décadi* as the day of rest. Days were to be composed of just ten hours (so 144 of our current minutes) and each hour was divided into a 100 minutes and each minute into 100 seconds. The whole reformed calendar lasted for twelve years, from 1793 to 1805, though the week and hour reforms never took off beyond the political periphery of Paris. It was revived for another eighteen days during the Paris Commune of 1871. It was ridiculed by the British, who nicknamed the Republican Calendar with its four formal seasons: Wheezy, Sneezy and Freezy; Slippy, Drippy and Nippy; Showery, Flowery and Bowery; Wheaty, Heaty and Sweety.

12 SYMBOLIC LINKS A BUDDHIST MUST BREAK TO ESCAPE SUFFERING

Avijja ❀ Sankhara ❀ Vinnana ❀ Nama-rupa ❀ Salayantana ❀ Phassa ❀ Vedana ❀ Tanha ❀ Upadana ❀ Bhava ❀ Jati ❀ Jara-marana

These are the twelve *nidanas* – the Buddhist concept of 'dependent origination'. Their cycle must be broken in order to escape suffering and attain nirvana.

Avija can be depicted as a blind old woman, representing unknowing – not knowing Buddhism's four noble truths. *Sankhara* is depicted as a potter representing volition – the forming of future lives by past karma. *Vinnana* (consciousness) is a monkey leaping from tree to tree, expressing our soul 'leaping' from life to life, formed by past karmic actions. *Nama-rupa* (mind–body) is depicted as a man in a boat, for the body and mind are the transport for our consciousness assisted by the six senses. *Salayantana* (the six senses) is a house with six windows, showing how our perceptions are gained through the portals of eye, ear, nose, tongue, touch and mind. *Phassa* (contact), a man and woman embracing, reminds us of contact between the six senses and external reality, which allows feelings to exist.

Vedana (feeling) is represented by a man with arrows through his eyes, for the emotional response to any contact falls into three categories of feeling – pleasant, painful and neutral. When feelings arise, cravings are produced. *Tanha* (craving)

The twelve nidanyas, represented as a wheel of life.

can be represented by a man drinking alcohol, for this craving leads towards greed and the making of bad karma. Only through being mindful can we understand and discipline our primal emotions.

By grasping and craving we have locked ourselves into a material existence and ensure more and more rebirths. *Upadana* (grasping) is depicted as a man reaching for fruit even though his basket is full. *Bhava* (becoming) is represented by a pregnant woman; *Jati* (birth), by a mother in labour. Where there is a birth, there is aways old age and death. *Jara-marana* (suffering) is represented by an old man with a corpse. Thus, as we live, we age and soon we will die. All things are impermanent. All things are suffering.

12 EMBLEMS OF SUPREME AUTHORITY

Sun ❀ Moon ❀ Stars ❀ Axe ❀ Fu ❀ Pair of sacrificial cups ❀ Water weed ❀ Mountains ❀ Five-clawed Dragon ❀ Pheasant ❀ Grain ❀ Fire

These twelve emblems were embroidered onto the Chinese Emperor's yellow silk robe during the Qianlong Period, the first six symbols on the front, the others on the back. The axe symbolised the power of execution; a Fu is a good fortune symbol; the dragon symbolised the power to guard from harm, rain, good fortune, full harvest and protection against fire; the grain the ability to feed; and the fire, brilliance.

12 SONS OF PAN

Kelaineus ❀ Argennon ❀ Aigikoros ❀ Eugeneios ❀ Omester ❀ Daphoineus ❀ Phobos ❀ Philamnos ❀ Xanthos ❀ Glaukos ❀ Argos ❀ Phorbas

The great god Pan – that randy, earthy, goat-footed deity of shepherds and wild places – always seemed set aside from the pantheon of the Olympians, like something left behind from a previous creation, touched with something ancient, primeval and Saturnine. The fact that he was worshipped in caves and grottoes seems to connect him with our most ancient holy places, and outside of that single shrine in Arcadia no great temples can be found raised to his worship, yet to an extent his cult survives, as the feared sensual other with which Christians dress up their image of Satan, no doubt reinforced by discovering classical statues of him tupping a she goat, chasing nymphs and teaching Hermes and Dionysus to enjoy acts of 'gross indecency'. So it is entirely suitable that his twelve sons all joined Dionysus (master of wine, intoxication, drama and sensuality) in his rollicking processional war into the East, spreading the culture of wine drinking and orgiastic dance, aided and abetted by Silenus.

12 FEASTS OF ORTHODOX CHRISTIANITY

Nativity of the Virgin ✳ Exaltation of the Cross ✳
Presentation of the Virgin to the Temple ✳ Christmas
✳ Epiphany ✳ Presentation of Jesus at the Temple ✳
Annunciation ✳ Palm Sunday ✳ Ascension ✳ Pentecost ✳
Transfiguration ✳ Dormition

This festival calendar was once utterly familiar across the Christian world – and it still shapes the Orthodox world. In fact, it is a vital checklist for anyone looking around an Orthodox church, trying to match their memory of the life of Christ to the gorgeous painted scenes. For here the feasts of the Church always have great prominence, as frescos painted high up on the walls above the ground-floor row of saints and in vivid scenes worked in wood, silver, mosaic or painted on gesso on the iconostasis. The twelve great feasts are of course capped by a thirteenth – the celebration of Easter itself, the triumphant apogee of the church year.

The Nativity of the Virgin (Theotokos), the birth of the Virgin Mary from Saint Anne, is celebrated on 8 September and symbolically begins the cycle of events, for the Virgin is the doorway for the incarnation of God in the form of Jesus into this world. Saint Anne was an elderly childless woman, who was simultaneously informed by the Holy Spirit that she was with child alongside her husband Saint Joachim, who was fasting in the desert at that time. The scene of them meeting at the Golden Gate of Jerusalem and falling into each other's arms, their eyes full of wonder and surprised happiness, is ever enchanting. The actual birth of the Virgin is usually represented with Saint Anna on a couch, assisted by two nursemaids about to wash her child.

The Nativity of the Virgin by Pietro Cavallini.

The *Exaltation of the Cross* celebrates, on 4 September, a series of events after the death of Jesus. This festival remembers the initial honouring of the Cross by Jesus's brother James as the instrument of our redemption and its rediscovery by Saint Helena (mother of Emperor Constantine) in 326 AD. Helena created the modern cult of the True Cross and the first Christian shrines and pilgrim rituals at Jerusalem, before the catastrophic sacking of the city by a Persian Sassanid army in 614. Fortunately it was miraculously found by Emperor Heraclius in 628; he carried it back to Jerusalem, walking barefoot on roads that had been covered with flowers by the ecstatic citizens of the Byzantine Holy Land.

The *Presentation of the Virgin (Theotokos)* is celebrated on 21 November, the day Mary is believed to have been presented by her parents to serve at the Temple as a young girl.

The *Nativity of Christ* is, of course, the birth of Jesus on Christmas Day. One common Byzantine detail is a midwife who doubts Mary's virginity and has her arm withered, and then restored after she bathes it in the baby Jesus's bathwater.

Epiphany is the acknowledgement of the divine within the body of Jesus. For most Christians this centres around the Baptism of Christ in Jordan by his cousin John, but it also references the visit of the Magi, and the water-into-wine miracle Jesus performed at the Marriage at Cana. All these events are celebrated together on 6 January.

The *Presentation of Jesus at the Temple* (the old English feast of Candlemas) is traditionally considered to have happened forty days after Jesus's birth (on 2 February) when his parents bought their newborn child to the Temple and offered a pair of turtle doves as a thanksgiving sacrifice (hence the Christmas song). On the steps of the Temple he was greeted by the mystic Simeon, who exclaimed the *Nunc Dimittis*, traditionally the final prayer of a church service.

The *Annunciation* depicts the Archangel Gabriel informing the Virgin Mary that she is pregnant with a miraculous child. This is always celebrated near to one of the traditional dates for the spring equinox, on 25 March. This event is attested in the Gospels of Luke and Matthew, and in the Koran.

Palm Sunday, the entry of Jesus on the back of a donkey into Jerusalem, is celebrated on the Sunday before Easter.

The *Ascension* is celebrated on a Thursday, forty days after the Resurrection that has been celebrated on Easter Sunday. At the Mount of Olives in the village of Bethany, Christ was observed to ascend into Heaven by eleven of his disciples – and this is believed to be the site of his second coming.

Pentecost (or *Whitsun* in the English traditional calendar) celebrates the gift of inspiration and courage (and fluency in the seventy tongues of mankind) descending on the gathered

Apostles (usually depicted as a white flame) exactly fifty days after Easter. They had met to celebrate the Jewish festival of Shavuot, the day on which Moses received the Torah.

The *Feast of the Transfiguration* is celebrated on 6 August and is the most spectacular and visionary image from the Orthodox artistic canon, depicting the body of Jesus transformed within a Heavenly mandala of clear, bright light (one of the Gospels writes of 'his face shining as the sun and his garments became white as light'). Saints Peter, John and James, who accompanied Jesus up the mountain (thought to be Mount Tabor), cower before the transfigured glory of their master as they observe him in conversation with the prophets Moses and Elijah and being addressed by a Heavenly voice as 'Son'.

The *Dormition of the Theotokos*, or the Assumption of Mary, is celebrated on 15 August. Imagery usually depicts Mary on her death-bed miraculously surrounded by eleven mournful Apostles, all in their canonical dress, seemingly unaware that Christ has descended in order to personally escort the soul of his mother (depicted as a miniature woman wrapped in a swaddling cloth) to Heaven.

LEAGUES OF 12 CITIES – DODECAPOLIS

Arezzo ❊ Cerveteri ❊ Chiusi ❊ Cortona ❊ Perugia ❊
Populonia ❊ Veii ❊ Tarquinia-Corneto ❊ Vetulonia ❊
Volterra ❊ Bolsena ❊ Volci

The Aeolian Greeks, Ionian Greeks and Etruscans all preserved ancient memories of how their ancestors were linked together in leagues of twelve that jointly honoured a sacred place. No definitive list survives, however. The cities above are the most widely accepted modern Italian cities that have an Etruscan foundation, though there may have been three separate Etruscan leagues, each composed of twelve cities.

12 PATRIARCHS

Shem ❊ Arphaxad ❊ Salah ❊ Heber ❊ Peleg ❊ Reu ❊
Serug ❊ Nahor ❊ Terah ❊ Abraham ❊ Isaac ❊ Jacob

With Noah as the sole possessor of the earth after the drowning of the earlier generations of mankind by the Flood, there is but a line of twelve patriarchal ancestors that take us from Noah to Jacob – himself the father of twelve patriarchal children who will found the twelve tribes of Israel.

Shem was one of the three sons of Noah (and is the ancestor of all the Semites of Arabia), his brothers establishing the peoples of Africa. *Salah* is much mentioned in the Koran. *Terah* was a sculptor who carved idols to the gods of Ur ,according to one tradition. *Abraham* fathered Ishmael to his Egyptian concubine Hagar – who would become the ancestor of all the Arab tribes, just as *Isaac*, the son of Abraham through his first Jewish wife, Sarah, would father *Jacob*, the ancestor of all the Jews.

12 TRIBES OF ISRAEL

Reuben ❊ Simeon ❊ Levi ❊ Judah ❊ Dan ❊ Naphtali ❊
Gad ❊ Asher ❊ Issachar ❊ Zebulun ❊ Joseph ❊ Benjamin

The twelve tribes of Israel are all descended from the twelve sons of Jacob – who was renamed Israel. They were not all from the same mother, for Jacob had children from his two wives as well as his two serving girls – and there was also a thirteenth descendant recorded, a much-overlooked daughter named Dinah.

Levi and his priestly descendants were given no land, while *Joseph* through his Egyptian wife Asenayh created two subtribes of Manasseh and Ephrim. The two tribes of *Judah* and

Benjamin united to form the Kingdom of Judah based on the city of Jerusalem, while the ten other tribes formed the northern kingdom of Israel with its separate capital of Samaria, which was to be utterly destroyed by the Babylonians.

THE 12 SHIITE IMAMS

Ali ibn Abu Talib ❀ Hasan ❀ Husayn ❀ Ali Zayn Al-Abidin ❀ Muhammad al-Baqir ❀ Jafar as Sadiq ❀ Musa Al-Kazim ❀ Ali er Rida ❀ Muhammad al Jawad ❀ Ali an-Naqi ❀ Hasan al-Askari ❀ Muhammad al-Mahdi

The Prophet Muhammad's cousin, son-in-law and first disciple, *Ali ibn Abu Talib*, is considered to be the first Imam by all Shia Muslims. He was blessed with spiritual authority in the last year of Muhammad's life at the caravan stopping place of Ghadir Khumm, though his place as rightful leader was taken first by Abu Bakr, then Omar, then Uthman before he ruled as Caliph.

After his assassination he was succeeded as rightful Imam by his son *Hasan*, followed by his younger brother *Husayn*, the martyr prince who was killed at Karbala. The only one of Husayn's family to survive was the young *Ali Zayn Al-Abidin*, who was so ill that he had been left behind in the camp. Then the line passes to his son *Muhammad al-Baqir* (5th Imam) and to his son *Jafar as Sadiq* (6th Imam).

The Ismaili Shia then trace the line descent from his son Ismail, while the majority of 'Twelver Shia' recognise his brother *Musa Al-Kazim* as 7th Imam, then down through *Ali er Rida* (8th Imam), *Muhammad al Jawad* (9th Imam) and *Ali an-Naqi* (10th Imam) to *Hasan al-Askari* (11th Imam), whose son *Muhammad al-Mahdi* is believed to be the 12th Imam who will come back again to usher in the end of time.

12 SONS OF ISHMAEL

Nebaioth ❖ Kedar ❖ Abdeel ❖ Mibsam ❖ Mishwa ❖
Dumah ❖ Massa ❖ Hadar ❖ Terna ❖ Jetur ❖
Naphish ❖ Kedemah

These twelve men are all the sons of Ishmael (who was the son of Abraham through his Egyptian concubine-wife Hagar). They are the patriarchal founders of all the tribes of Arabia. The Quryash of Mecca (a confederation of clans from whom the Prophet Muhammad was descended) trace their descent from Ishmael's second son, *Kedar*. The sons of *Massa* settled the good grazing lands on the western-bank of the Tigris, in modern Iraq. There is also a belief that after the death of his wife Sarah, Abraham came to live with Hagar in the Arabian desert and she gave birth to *Keturah*, a much young brother to Ishmael and ancestor of all the tribes of Syrian Arabs.

12 PRECIOUS STONES

Sardius ❖ Topaz ❖ Emerald ❖ Turquoise ❖ Sapphire ❖
Diamond ❖ Jacinth ❖ Agate ❖ Amethyst ❖ Beryl ❖
Onyx ❖ Jasper

These twelve precious stones have become strongly associated with the twelve months of the year, the tribes of Israel, and the signs of the zodiac. There is no orthodoxy as to which jewel should be linked with which month, though Exodus 28:15-21 asserts: 'You shall make the breastplate of judgment. Artistically woven according to the workmanship of the ephod you shall make it: of gold, blue, purple, and scarlet thread, and fine woven linen, you shall make it. It shall be doubled into a square: a span shall be its length, and a span shall be its width. And you shall put settings of stones in it, four rows of stones: The first row shall be a sardius, a topaz, and an emerald; this shall be the first row; the second row shall be a turquoise, a sapphire, and a diamond; the third row, a jacinth, an agate, and an amethyst; and the fourth row, a beryl, an onyx, and a jasper. They shall be set in gold settings. And the stones shall have the names of the sons of Israel, twelve according to their names, like the engravings of a signet.'

12 CHINESE CHARACTER YEARS

Rat (1924, 1936, 1948, 1960, 1972, 1984, 1996, 2008) ❊ Ox (1925, 1937, 1949, 1961, 1973, 1985, 1997, 2009) ❊ Tiger (1926, 1938, 1950, 1962, 1974, 1986, 1998, 2010) ❊ Hare (1927, 1939, 1951, 1963, 1975, 1987, 1999, 2011) ❊ Dragon (1928, 1940, 1952, 1964, 1976, 1988, 2000, 2012) ❊ Snake (1929, 1941, 1953, 1965, 1977, 1989, 2001, 2013) ❊ Horse (1930, 1942, 1954, 1966, 1978, 1990, 2002, 2014) ❊ Sheep (1931, 1943, 1955, 1967, 1979, 1991, 2003, 2015) ❊ Monkey (1932, 1944, 1956, 1968, 1980, 1992, 2004, 2016) ❊ Cock (1933, 1945, 1957, 1969, 1981, 1993, 2005, 2017) ❊ Dog (1934, 1946, 1958, 1970, 1982, 1994, 2006, 2018) ❊ Pig (1935, 1947, 1959, 1971, 1983, 1995, 2007, 2019)

Throughout China, Japan and Central Asia, births are placed within a recurring cycle of symbolic character animals. Above are birth years (from 1924–2019) and here are their chief characteristics:

Rats have spirit, wit, alertness, delicacy, flexibility and vitality. They are smart, accumulate wealth and make efforts to be successful. They are loving to their intimates and respect family life, but must keep a lid on their desire to win.

Oxes are sedulous, simple, honest and straightforward. They are laborious and patient but can also be obstinate and poor at communication. Female oxes are usually good wives who pay attention to children's education but are often credulous. Ox-people, however, tend to age well.

Tigers are brave and forceful yet also tolerant, staunch, valiant and respected. In their middle age, they must avoid becoming cruel and bullying out of thwarted ambition, but in old age these faults should mellow. Women tigers tend to be happier than men and are intelligent, faithful and virtuous.

Hares are tender, lovely and full of hope. They are gentle, sensitive, modest and merciful and have strong memories. They like to communicate with others in a humorous manner and cannot bear a dull life, so though good at creating interesting events they can also devise hare-brained schemes.

Dragons have a capacity for dignity, honour, success and luck. They are lively and energetic but prone to fight over leadership, perfectionism, and can be impatient and arrogant.

Chinese character zodiac from a seventh-century Tang dynasty mirror.

Snakes have a good temper, skill in communicating, and morality, but a tendency to jealousy and suspicion. They must take care in discussing others.

Horses are energetic, bright, warm-hearted, intelligent, active, clever, kind to others and adventurous. But they cannot bear too much constraint and prefer surface tactics to depth.

Sheep or goats are gentle, popular and calm, tender, polite, filial, clever and kind-hearted. They are sensitive to art and beauty, faith in religion and like quiet, economical living. Women take good care of others, but should struggle to avoid too much hesitation and self-doubt.

Monkeys are clever animals – lively, flexible and versatile. They love sports, activity, communicating and helping others. In work they do not like to be controlled, but can have an amazing creativity in the right environment, just as in the wrong one they can be critical, impatient and mouthy.

Cocks are the epitome of fidelity and punctuality, honest, bright, communicative, ambitious and warm-hearted. They have strong self-respect and seldom rely on others, but have a tendency to sudden impulses driven by arrogance. They are often attractive and might have several loves in their lives.

Dogs are auspicious animals and have a straightforward character. In their career and love, they are faithful, courageous, dextrous, smart and warm-hearted. However, they must watch out for a lack of stability.

Pigs are enthusiastic, extroverted, rebellious, passionate, brave and valiant; however, they can also be hot-tempered, snappy, uncontrollable and short-tempered.

11

THE 11TH HOUR

Significant 11s punctuate modern history. The First World War, after consuming some twenty million lives, ended with an armistice on the 11th hour of the 11th day of the 11th month. Eleven is also strongly associated with America and rocket power, for it was Apollo 11 from which Neil Armstrong made the first landing on the moon. The attack on the World Trade Centre was made by American Airlines flight 11, on 11 September 2001, an event now chronicled throughout the world as 9/11. Eleven also has strong numerological connotations as the union of 5 and 6 in the works of Pythagoras and his many followers.

DANTE'S 11

Dante was a keen follower of Pythagoras, the sixth-century BC Greek philosopher and mathematician who sought to explain the world, both spiritual and material, by numbers. Pythagoras believed that the mathematical principles that underlay the universe gave it a harmony, literally a music of the spheres. Dante, in his great work the *Divine Comedy*, sought to create the divine song.

The key number for Dante was 11 – the union of 5 and 6 – and its multiples. The *Inferno*, *Purgatorio* and *Paradiso* have thirty-three cantos each, and the poem is written in hendecasyllabic rhyme (eleven syllables long). Dante twice provides dimensions for Hell, stating that the circumference of the ninth bolgia (ditch) in the Eighth Circle is 22 miles (miglia ventidue), and the tenth bolgia is 11 miles. There is nothing accidental about this mention of 11 and its multiple 22; twenty-two forms part of the well-known fraction 22/7 which expresses the Pythagorean value of pi.

Three and nine also figure prominently in Dante's numerology. The three books of the *Divine Comedy* delineate the nine circles of Hell, the nine rings of Mount Purgatory and the nine celestial bodies of Paradise.

....

11 FOOTBALLERS

....

Historians of numbers have long noted that eleven makes a team: be it the eleven-strong team of detectives charged with identifying murderers in ancient Rome, the eleven strong vice squad of women established by the Spartan state to keep a lid on Dionysiac orgies or the eleven monstrous giant-beasts who assisted Tiamat, Lord of Chaos in his doomed battle against the pantheon of Babylon led by the King of the Gods, Marduk.

Of course, the main 11 in modern life is that of football teams. The eleven traditional positions are: goalkeeper, left back, right back, left half, centre-half and right half, and the five forward positions – left wing, inside left, centre-forward, inside right and right wing. However, as illustrated by the classic line-up opposite – the team that won England's only World Cup title (in 1966) – the formations have long had different permutations, most commonly with the outfield players arranged as 4–3–3 or 4–4–2.

You would imagine that the 11 players in a football team must have some significance, for the game is as old as humanity, but in its early centuries football was often played by hundreds of people. The game as we know it today was developed largely in the English public schools of the nineteenth century and first codified in the 'Cambridge Rules' of 1848, thrashed out from the various schools different traditions. Oddly, there

England's World Cup formation against Beckenbauer's West Germany.

was no mention of the number of players in a team then, nor in the earliest surviving set of rules from 1856, though the number had been accepted by the time the FA Cup was established in 1871. It may have followed the example of cricket (see below), with a convenient division of five attackers, five defenders and a goalkeeper.

Rugby, of course, went its own way in the nineteenth century, allowing not just the handling and carrying of the ball but a player count of 13 or 15.

CRICKET'S FIRST 11

Eleven a side seems to have become the standard for cricket very early on in the game, for reasons equally as obscure as in football. But of course it, too, makes sense: five decent bats, five bowlers and a wicketkeeper. The earliest report of a cricket match was in the *Foreign Post* of 7 July 1697 and describes a match played in 'the middle of last week in Sussex with eleven of a side and they played for fifty guineas apiece'.

$\underline{10}$

---

2 HANDS – 10 FINGERS

The prime motivation behind the power of 10 is that you can with some authority recite your list of laws, prophets or gods as you tick of each of your ten fingers from a pair of hands. So the decision to decimate a rebel legion, to take a tithe of a tenth of the harvest as tax or to rule for a decade seems logical, absolute and ordained. The decimal system which now rules our numerical world, our wealth, our conception of time and distance derives from *dekm* – the Indo-Aryan word for 'two hands', the power of ten.

---

10 PLAGUES OF EGYPT

Water Turned to Blood ❖ Plague of Frogs ❖ Plague of Ticks ❖ Plague of Flying Insects ❖ Cattle Disease ❖ Boils ❖ Great Hail ❖ Plague of Locusts ❖ Thick Darkness Death of all the Firstborn in the Land of Egypt

The plagues of Egypt are recounted in Exodus and repeated in the Koran. The last of the afflictions called upon the people of Egypt by Moses is the most chilling, for the Israelites are only spared from the attention of the angel of death by smearing their doorposts with the blood of the lamb they have just sacrificed for the feast of the Passover.

THE 10 COMMANDMENTS

Thou shalt have no other gods ✳ Thou shalt not make any graven images ✳ Thou shalt not take the Lord's name in vain ✳ Remember the Sabbath day ✳ Honour thy father and mother ✳ Thou shalt not kill ✳ Thou shalt not commit adultery ✳ Thou shall not steal ✳ Thou shall not bear false witness against thy neighbour ✳ Thou shalt not covet thy neighbour's house, nor his oxen, nor anything that is his

The Ten Commandments were first spoken by God to the terrified people of Israel and then reaffirmed when Moses brought down from the cloud-obscured summit of Mount Sinai two stone tablets inscribed by the finger of the Almighty.

They are listed twice in the Old Testament, so can have minor variations in the text, but there are no essential contradictions if you compare Exodus 20: 1–17 with Deuteronomy 5:4–21. Although often depicted in synagogues and churches or ornamental texts as two tablets side by side, the tradition is that the two stone tablets were identical, like a signed and sealed treaty accompanied by an exact duplicate.

10 GURUS OF THE SIKHS

Guru Nanak ❋ Guru Angad ❋ Guru Amar Das ❋
Guru Ram Das ❋ Guru Arjan ❋ Guru Har Gobind
❋ Guru Har Raji ❋ Guru Har Krishan ❋ Guru Tegh
Bahadur ❋ Guru Gobind Singh

The ten gurus are historical figures within the Punjab, but they are also conceived as one spirit, reincarnated in ten lives. The era of the ten historical Gurus lasted for 239 years, beginning in 1469 with Guru Nanak and ending with the death of the tenth Guru Gobind Singh in 1708, after which their collected hymns, teachings and scripture reigns as the eternal Guru Granth Sahib.

10 MAGPIES

One for sorrow ❋ Two for joy ❋ Three for a wedding ❋
Four for a death ❋ Five for silver ❋ Six for gold ❋ Seven
for a secret not to be told ❋ Eight for Heaven ❋ Nine for
Hell ❋ And ten for the devil coming for your soul

It is tempting to see the counting of magpies, and the chanting of verses about what the number of these distinctive black and white birds might mean, as a tenuous but still active strand of traditional lore that links us directly to the ancient art of divination by watching the flight of birds (augury). We know that priests throughout the ancient world attempted to read the future by watching the passage of birds pass some sacred feature, such as a temple sanctuary, a headland or the gates of a city. The direction of their flight, their species, their number, the month, the hour and the shape of the flocks must all have had a significance that is now lost to us. One has only to think of the shapes formed by starlings at dusk, the vast squadrons of migrating geese or gulls returning every night to the sea, to touch upon the complexities of this art, let alone what the chance sighting of an erratic might configure ...

The counting of magpies encountered on an English pathway drops us into a more homely version of this lost science, concerned with the fate of an immediate family. As with any ancient oral tradition, there is a considerable variety after

One for sorrow? Piero della Francesca painted a single magpie at the apex of his unfinished masterpiece of the Nativity.

the first two lines. The most popular version in use today stops at seven. It is also worth noting that the magpie is a meat-eating corvid, and that the British have always tended to weave gloomy references around magpies and their bigger cousins, such as crows and most especially the raven. All these birds would have been seen to feed off the dead of the battlefield, or those left swinging on the gallows, exhibited on the gibbet or impaled on a pike.

10 PAINS OF DEATH

To wait for one who never comes ❊ To lie in bed and not to sleep ❊ To serve well and not to please ❊ To have a horse that will not go ❊ To be sick and lack the cure ❊ To be a prisoner without hope ❊ To lose the way when you would journey ❊ To stand at a door that none will open ❊ To have a friend who would betray you

'These are the ten pains of death' ennumerated in a sixteenth-century poem, *Second Fruits* by Giovanni Florio, as quoted in Gavin Maxwell's study of western Sicily, which he called *The Ten Pains of Death*. Maxwell's own book circles around the plots and political conspiracy of post-war Sicily and the glamorous Mafiosi bandit, Salvatore Giuliano. The poem has a bitter bearing on Maxwell's study, for Giuliano was almost certainly betrayed by his friends.

10 DEITIES WITHIN BES

Aha ❊ Amam ❊ Bes ❊ Hayet ❊ Ihty ❊ Mefdjet ❊ Menew ❊ Segeb ❊ Sopdu ❊ Tetenu

Bes is a fascinating deity: a diminutive, bandy-legged and bearded dwarf with an ugly mask-like face, big ears, large staring eyes and a dangling penis. He, or rather his ten-fold

incarnations, is usually depicted with a protruding belly over which sags a pair of corpulent middle-aged male breasts, and can be clad in a grass kilt and made almost ridiculous by wearing an oversize crown of ostrich feathers. Yet this god was everywhere in the ancient world – depicted on a Persian golden chariot, or as a carved figurehead on the dark ships of the Phoenicians, in profile guarding tombs and temples, and carved into thousands upon thousands of amulets. For Bes was a valuable god to have on your side. He protected women in childbirth, watched over the make-up table and drains, and seems to have been helpful in sexual matters. Though there is a vast statue of him discovered in Amathus in Cyprus, he usually is not given a temple of his own, but like some ancient much-loved heroic earth spirit is everywhere. His name derives from *besa*, 'to protect'.

10 SEFIRAH OF THE KABBALAH

Chokhmah ❋ Binah ❋ Daat ❋ Chessed ❋ Gevurah ❋ Tiferet ❋ Netzach ❋ Hod ❋ Yesod ❋ Malkuth

These ten emblems have been used as tools to understand the inner constitution of all reality as well as manifestations from the divinity that sustains mankind and all nature. *Chokhmah* represents Wisdom and Conception; *Binah*, understanding and comprehension; *Daat*, knowledge and attachment; *Chessed*, kindness, loving and giving; *Gevurah*, strength, restraint and discipline; *Tiferet*, beauty, harmony and empathy; *Netzach*, ambition and fortitude; *Hod*, devotion and humility; *Yesod*, foundation and the bonding link between body and mind; *Malkuth*, royalty and the world of the body and its energies. *Ketter* – the crown, our subconscious – is sometimes used to replace Daat.

They are highly fluid, interpreted anew by each generation, yet of ancient tradition, emerging as elements of Jewish mysticism in the third century BC. They seem to have first found written form in medieval Spain and would be further developed by savants in the city of Safed in ancient Palestine.

They are closely linked to the belief in the Ten Attributes of God as revealed when the Absolute All withdrew his presence in order to behold himself. These, from highest to lowest, are the powers of: Supreme Consciousness, Wisdom, Love, Vision, Intention, Creativity, Contemplation, Surrender/Sincerity, Memory/Knowedge and Healing.

10 QUALITIES OF A BEDOUIN WARRIOR

Courage of a cock ❈ Painstaking nature of a chicken ❈ Heart of a lion ❈ Brusqueness of a boar ❈ Tricks of a fox ❈ Prudence of a hedgehog ❈ Swiftness of a wolf ❈ Resignation of a dog ❈ A hand always open ❈ A sword always drawn ❈ One speech for friend and foe

If the evidence of the pre-Islamic verses are anything to go by, the Bedouin were also renowned lovers. However, according to a traditional saying of Imam Ali, this was as nothing in comparison to the women of Arabia – for 'God created sexual desire in ten portions, then he gave nine parts to women and one to man.'

9
=

A PANTHEON OF 9

In the Indo-Aryan West, 9 was always a most propitious number, for each aspect of a ruling trinity could be multiplied by 3 to create a pantheon of 9. It is not difficult to see that this was the likely origin for our choirs of nine angels, nine Heavenly spirits and nine muses. The number also has an innate reference to the nine months in which a child is created within the womb.

9 CHOIRS OF ANGELS

Seraphim ❊ Cherubim ❊ Thrones or Ophanim ❊
Dominions ❊ Virtues ❊ Powers ❊ Principalities ❊
Archangels ❊ Angels

The nine orders of the Celestial Hierarchy of Angels were established for the early Christian church by Saint Jerome and Pseudo-Dionysis the Areopagite in the fourth century. The Second Council of Nicaea (787) provided formal permission for this depiction.

The lowest order of angels is that of Messengers, composed of *Angels*, *Archangels* and *Principalities*. The second order are the governors of stars and elements, ranked as *Powers*, *Virtues* and *Dominions*; in iconography they are coloured blue for light and knowledge. The third and highest order are Counsellors, arranged in three ranks: *Thrones*, *Cherubim* and *Seraphim*. They are coloured red for love, with the two highest orders just depicted with the heads of pure intelligences. A seraph of the choir of Seraphim, such as Uriel, is coloured crimson for burning love and is depicted with six wings. Two wings cover the face, two cover the feet and two are used for flying. They sing praises around the throne of God and carry scrolls saying, 'Holy, Holy, Holy is the Lord of Hosts.'

The first Christian angels (from the Greek *angelos*) were seemingly modelled on the classical depictions of Cupid and Nike-winged victory. They also followed Jewish tradition, which had ten ranks, according to the great Sephardic Renaissance man Moses Maiomonides.

9 MUSES

Clio ❋ Euterpe ❋ Thalia ❋ Melpomene ❋ Terpsichore ❋ Erato ❋ Urania ❋ Calliope ❋ Polyhymnia

The nine muses, the daughters of Zeus and Mnemosyne (the goddess of memory), were a favourite subject for Roman artists and much depicted in mosaic and fresco, or carved in marble to grace the praesidium of a theatre.

Clio, the muse of history, is represented with a stylus and a scroll, or after the Renaissance with a book, a laurel crown or a trumpet; she is easy to confuse with *Calliope*, who often has the same attributes. *Euterpe*, muse of lyrical poetry, bears a flute. *Thalia*, muse of pastoral poetry and comedy, carries a comic mask and sometimes a viol.

Melpomene, muse of tragedy, is associated with a mask, sometimes embellished with a fallen crown, and holds a dagger. *Terpsichore,* muse of joyful dance and song, often holds a lyre, as does *Erato,* muse of lyrical love poetry.

Urania, muse of astronomy, is normally shown consulting a globe or a compass. *Polymnia,* muse of heroic hymn and eloquence, possesses a lute and a solemn expression that outdoes even those of Clio and Calliope.

Clio – or just possibly Calliope.

9 FIRE ALTARS OF VICTORY

Since the Islamic suppression of Zoroastrianism in its homeland of Iran, just nine temples were left to maintain the Atash Behram – the Fire of Victory that must be continuously tended. The Atash Behram is the third and highest grade of fire, above the Atash Dadgah and the Atash Adaran, and can only be created from merging sixteen different sources of fire (including that incubated by a lightning bolt) in a long ceremony that requires the participation of thirty-two priests. Eight of the nine altars are now located in India, though one remains in the Iranian homeland, at Yazd, where it was inaugurated by a Sassanian Shah in 470 AD.

The symbolism of the number 9 embedded in the number of Atash Behram evolved over the last couple of hundred years but seems well established. The number is also manifest in the nine priestly families of Sanjan who collectively form a high priesthood, as well as the Zoroastrian belief in the ninth day of the ninth month as propitious.

9 NIGHTS OF ODIN'S SACRIFICE

Odin, the chief Norse god, made a sacrifice to himself, plucking out one eye and hanging for nine days and nine nights from the world tree Yggdrasil, pierced through his side by his magical spear, Gunghir. This allowed his soul to wander and gain insight into the the nine realms of existence as well as to learn two sets of nine magical songs and rune spells. This shamanic sacrifice is told in the Norse *Havanal* epic: 'Downwards I peered; I took up the runes, screaming I took them, then I fell back from there.'

9 AZTEC LORDS OF THE NIGHT

Xiuhtecuhtli (Turquoise Lord) ❋ Tecpati-Itztli (Lord of the Obsidian Blade) ❋ Piltzintecuhtli (Our Lord Prince) ❋ Centeotl (Lord of the Maize) ❋ Mictlantecuhtli (Underworld Lord) ❋ Chalchiuhtlicue (Lord of the Jade skirt) ❋ Tlazoteotl (Our Lady of Two Faces – lustful sin and purification) ❋ Tepeyollotl (Lord of the Heart of the Mountain) ❋ Tlaloc (Lord of Rain and Fertility)

The Aztecs, like most of the pre-Columbian civilisations of Meso-America, ran a number of sacred calendars concurrently, which made life much more interesting, in terms of working out festivals and celebrations, as well as good, bad, propitious and impossible days, nights and months. Blocks of nine nights fitted into both the 365-day-long solar year (known as Haab), which was divided into twenty groups of

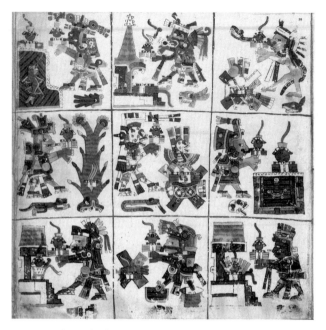

The Lords of the Night depicted in the Codex Borgia:
(1a) Tlaloc, (1b) Tepeyollotl, (1c) Tlazoteotl, (2a) Centeotl,
(2b) Mictlantecuhtli, (2c) Chalchiuhtlicue,
(3a) Piltzintecuhtli, (3b) Tecpati/Itztli, (3c) Xiuhtecuhtli.

eighteen-day months, as well as the 260-day-long fertility calendar (known as Tzolkin) composed of twenty groups of thir-
teen-day months as well as twenty-nine groups of nine nights.

Twenty-nine is of course the unit of a lunar month, while
nine months represent the gestation of both a human child
and the complete tropical cycle of sowing to reaping for
such vital crops as maize. So the Lords of the Night, in
some South American cultures, appear also as the Lords of
the Nine Months, or the Nine Judges of Hell, and in other
ninefold manifestations.

9 MEXICAN POSADAS

The nine Mexican Posadas are a series of dances, candle-lit processions, recitals and songs held over the nine nights before Christmas. They tell the story of the Holy Family (the pregnant Mary and Joseph) travelling out of Galilee to Judaea to try and reach Bethlehem. On the last night, Joseph once again sings his desperate refrain to an empty door – 'the night is cold and dark and the wind blows hard' – before Mary accidentally reveals that beneath their travel-worn cloaks she is Queen of Heaven and she is welcomed into a stable by the animals. Then a dance of honour is held and a 'piñata' is demolished by a blindfold young lady wielding a cane to shower sweets over the celebrants.

9 RASA

Srngaram ❖ Hasyam ❖ Raudram ❖ Karunyam ❖ Bibhatsam ❖ Bhayanakam ❖ Viram ❖ Abdhutam ❖ Santam

The Nine Rasa are a constant cultural refrain underlying Indian popular art, music, theatre, comics and cinema. They are noisy, plebeian and fun-loving.

Srngaram (love, attractiveness and the erotic) is presided over by Lord Vishnu and coloured light green. *Hasyam* (laughter, mirth and comedy) is presided over by Pramata and coloured white. *Raudram* (fury) is presided over by Rudra and coloured red. *Karunyam* (compassion and tragedy) has Yama as its presiding deity and is coloured grey. *Bibhatsam* (disgust, aversion and the pathetic) is presided over by Shiva and coloured blue. *Bhayanakam* (horror) is presided over by Kali and coloured black. *Viram* (heroic endeavour) has as its presiding deity Indra, and is coloured yellow. *Abdhutam* (wonder and amazement) has Lord Brahma and is coloured bright yellow. *Santam* (peace and tranquillity) has the presiding deity of Vishnu and is coloured blue.

9 SKILLS OF A NORSE LORD

Chessplayer ❋ Runeteller ❋ Bookreader ❋ Ironworker ❋ Skier ❋ Archer ❋ Rower ❋ Musician ❋ Poet

As Earl Rognvald of Orkney in the *Orkenyinga Saga* (1200 AD) tells us: 'I can play at Tafl, Nine skills I know, Rarely I forget the Runes, I know of Books and Smithing, I know how to slide on skis, Shoot and row well enough, Each of the Two arts I know, Harp playing and speaking poetry.'

9 FREEDOMS OF THE ENGLISH DECLARATION OF RIGHTS

No royal interference in the laws ❋ No taxation without Parliament ❋ Freedom to petition without fear ❋ No standing army without popular consent ❋ The people have the right to bear arms for their own protection ❋ No royal interference in elections ❋ Total freedom of speech in Parliament ❋ No grant or promise before a trial can have any validity ❋ No cruel and unusual punishments

Though it has none of the grandeur of the French Rights of Man and the Citizen, or the tectonic universalism of the American Declaration of Independence and the Bill of Rights, the English Declaration of Rights of 1688/89 is yet the proud father to these later and more famous assertions of political freedom.

For the rest of the world there is too much British dynastic politics in the creed and its additional clauses, which declared that a papist prince is incompatible with the safety and welfare of a Protestant kingdom, and that a king (such as James II) in exiling himself to a foreign power had abdicated and settled the immediate line of succession in a Protestant line of descent. The Declaration was tactfully presented to William and Mary but not made a precondition of their coronation and became the Bill of Rights the following year.

9 WORTHIES (*LES NEUF PREUX*)

Hector, son of Priam of Troy ❊ Julius Caesar ❊ Alexander the Great ❊ Joshua ❊ King David ❊ Judas Maccabeus ❊ King Arthur ❊ Emperor Charlemagne ❊ Godfrey de Bouillon

This grouping of heroes – three pagans, three Jews and three Christians – formed an international pantheon of chivalry. The list was first recorded in *Les Voeux du Paon* by Jacques de Longuyon at the beginning of the fourteenth century, whereafter it became a popular motif in ballads, romance literature, tapestries and pageants. For instance, when Henry VI entered Paris in 1431 he was preceded by the Nine Worthies, set off by Nine Heroines. French kings liked to pose as the Tenth Worthy.

9 PERSONALITIES OF THE ENNEAGRAM

Perfectionist ❊ Giver ❊ Achiever ❊ Tragic/Romantic ❊ Observer ❊ Contradictor ❊ Enthusiast ❊ Leader ❊ Mediator

The nine-sided enneagram was popularised by the Greek-Armenian spiritual teacher George Gurdjieff (1866–1949) as a way of both analysing and then reforming character – for the final goal is to achieve a balance of all these aspects. It is also associated with the Kabbalah, while the attributes are close to those of the ideal god-loving Muslim, who is both

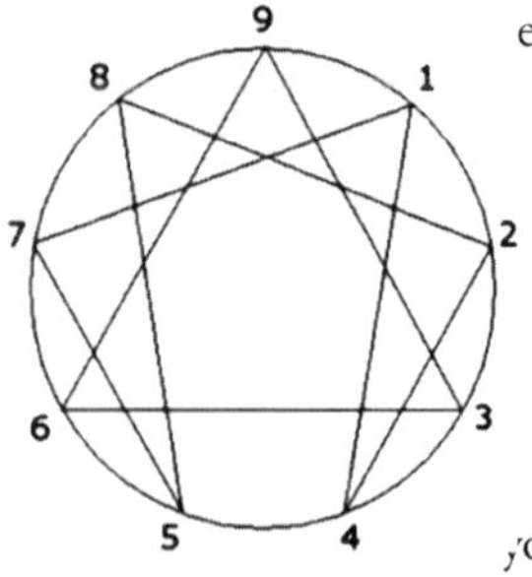

eformer, helper, achiever, individualist, investigator, loyalist, enthusiast, challenger and peacemaker.

Gurdijeff's enneagram was developed by the psychologists Oscar Ichazo and Claudio Naranjo to create a personal character map, which you can do easily enough for yourself. By marking your characteristics (on, say, a scale of 1 to 10) you can chart your particular strengths and weaknesses, and then see how these compared to those drawn for you by your friends or a counsellor.

CHINESE POWER OF 9

Nine has always been respected by the Chinese, for it has tonal resonance with 'long lasting' and was also associated with the Emperor, who had nine dragons embroidered on his robes and ruled over a court divided between nine ranks of courtiers who could gain nine sorts of reward. This respect for the power of 9 led to many social listings of 9, often charged with an observant sense of humour, as well as the more serious concept of how individuals were bound ninefold to their family, clan and community.

Here are the 9 Admirable Social Habits:

Relieving tension ❊ Courteous attention ❊ Discreet mention ❊ Tenacious retention ❊ Assiduousness ❊ Wise abstention ❊ Calculated prevention ❊ Truthful intervention ❊ A sense of dimension

The 9 Virtues – as defined for the near legendary Emperor Yu (2205–2100 BC) by his chief minister Kao-Yao:

Affability combined with dignity ❋ Mildness with firmness ❋ Bluntness with respectfulness ❋ Ability with reverence ❋ Docility with boldness ❋ Straightforwardness with gentleness ❋ Easiness with discrimination ❋ Vigour with sincerity ❋ Valour with goodness

The 9 Follies:

To think oneself immortal ❋ To think investments are secure ❋ To mistake conventional good manners for friendship ❋ To expect any reward for doing right ❋ To imagine the rich regard you as an equal ❋ To continue to drink after you have begun to declare that you are sober ❋ To recite your own verse ❋ To lend money and expect its return ❋ To travel with too much luggage

The 9 Jollities of a Peasant:

To laugh ❋ To fight ❋ To fill the stomach ❋ To forget ❋ To sing ❋ To take vengeance ❋ To discuss ❋ To boast ❋ To fall asleep

The 9 Deplorable Public Habits

Drunkenness ❋ Dirtiness ❋ Shuffling ❋ Over-loud voice ❋ Scratching ❋ Unpunctuality ❋ Peevishness ❋ Spitting ❋ Repeated jests

And the 9 Final Griefs:

Dissapointed expectations ❋ Irretrievable loss ❋ Inevitable fatigue ❋ Unanswered prayers ❋ Unrequited service ❋ Ineradicable doubt ❋ Perpetual dereliction ❋ Death ❋ Judgement

8
===

8 CHERRY STONES

Tinker ❊ Tailor ❊ Soldier ❊ Sailor ❊ Richman ❊
Poorman ❊ Beggarman ❊ Thief

This childhood rhyme helps make the eating of a thick slice of cherry tart, or a bag of fruit, an even more enjoyable task. For you hoard your cherry stones at the side of the plate, which, once you have finished, are counted out to an eight-fold repeating rhythm to reveal your future career, or that of your partner.

In England it normally goes as above, though I have heard variations like 'Tinker, Tailor, Soldier, Sailor, Gentleman, Apothecary, Plough-boy, Thief', or the Scots version, which goes, 'A Laird, a Lord, a Richman, a Thief, a Tailor, a Tinker, a Drummer-boy, a Stealer o'beef.' In America, 'thief' has been replaced with 'Indian chief' and the invidious choice between 'Richman' and 'Poorman' by a choice of the pro-fessions, 'Doctor, Lawyer or Merchant'. My mother used to chant out a genteel English version of this winners-eight, where all future partners could be considered gentlemen. It went: 'Soldier Brave, Sailor True, Skilled Physician, Oxford Blue, Gouty Nobleman, Squire-so-Hale, Dashing Airman, Curate Pale.'

Writers have been consistently drawn to the fateful roll call of destiny, either including the rhythm in their works or playing around with variations. So there are references to the oral chant to be found in titles and works by Thomas Hardy, Dorothy Sayers, Virginia Woolf, Michael Ondaatje, William Congreve, A.A. Milne and, most famously, John Le Carré.

One of the more intriguing references to a list of eight is also the earliest. In William Caxtons's edition of *The Game and Playe of the Chesse*, made around 1475, he names the eight pawns on the chess board as 'Labourer, Smith, Clerk, Merchant, Physician, Taverner, Guard and Ribald' – which might just be the origin of the whole chant.

8 TRIGRAMS OF THE I-CHING

Ch'ien ❋ Tui ❋ Li ❋ Chen ❋ Sun ❋ K'an ❋ Ken ❋ Kun

The I-Ching or Zhouyi, or the *Book of Changes* is the oldest of the Chinese classics, going back to oral traditions and observations of mankind at least 4,000 years old. It is essentially a collection of six-line hexagrams which are arranged in a textual eightfold pattern, which come with a set of linked values, such as an image in nature, a compass direction and an associated animal. By the use of chance (the casting of coins, dice, yarrow stalks or whatever), the sequences can be changed so that different groups of six-line hexagrams are read together, which gives it the force of a horoscope, managing questions with a cryptic and ever-changing set of responses.

The eight trigrams each have an association with a form of male or female energy, a place, a direction of the compass and a characteristic animal. For instance, the *Ch'ien* trigram is associated with creative force, with Heaven, the north-west and the horse; the *Tui,* with joyous openness, lake, west, sheep; the *Li,* with beauty and radiant awareness, fire, south and pheasant; *Chen,* with action and movement, thunder, east and dragon; *Sun,* with following and penetration, with wind, the south-east and the fowl; *K'an,* with danger and peril, water, north and the pig; *Ken,* with stopping and resting still, with mountains, the north-east and the wolf/dog; *Kun,* with receptive, earth, southwest and the cow.

The eight trigrams of the I-Ching.

8 IMMORTALS

Chung-li Ch'uan ❋ Ho Hsien-ku ❋ Chang Kuo ❋
Lu Tung-pin ❋ Han Hsiang-tzu ❋ Ts'ao Kuo-chiu ❋
Li T'ieh-kuai ❋ Lan Ts'ai-ho

This Taoist pantheon of gods, heroes and historical individuals had by the thirteenth century become a sort of national pantheon of Chinese saints. Painted on silk, depicted on vases, sculpted and used as a central motif in story telling, they are a ubiquitous element in art. They are also known as the Eight Immortal Scholars of the Han.

Chung-li Ch'uan is usually depicted as a bearded sage with fan; *Ho Hsien-ku,* as a young girl holding a lotus; *Chang Kuo* is a comical bearded figure mounted back to front on a white mule with a bamboo drum; *Lu Tung-pin*, the bearded patron of barbers, is equipped with a fly whisk and sword slung across his back; *Han Hsiang-tzu* is a youthful flute player and the patron saint of musicians; *Ts'ao Kuo-chiu* is an elderly bearded figure (the patron of actors) usually seen playing castanets; *Li T'ieh-kuai* is a beggar with gourd-bowl and iron crutch; while *Lan Ts'ai-ho* is a woman holding a basket of flowers, who is (naturally) the patron saint of florists.

EIGHTFOLD PATH OF THE BUDDHA

Right View ❋ Right Intention ❋ Right Speech ❋
Right Action ❋ Right Livelihood ❋ Right Effort ❋
Right Mindfulness ❋ Right Concentration

This is not a sequential course of study that is to be ticked off with examinations and advancements to the next stage of the path, but a path to be engaged with all your life. It must begin with a clear view of the Four Noble Truths (see *4*).

ARISTOTLE'S EIGHTFOLD CHAIN OF POLITICS

The world is a garden fenced by the state ⁕ The state is
power supported by law ⁕ Law is the policy which guides
the ruler ⁕ The ruler is order protected by the army ⁕
The army are supporters sustained by money ⁕ Money is
sustenance produced by subjects ⁕ Subjects are servants
protected by justice ⁕ Justice is ingrained and is the
support of the world

Aristotle's Circle of Politics was elaborated by Ibn Khaldun
to create the eight self-sustaining links in the chain of medieval statecraft.

EIGHTFOLD PATH OF YOGA

Yama ⁕ Niyama ⁕ Asana ⁕ Pranayama ⁕
Pratyahara ⁕ Dharana ⁕ Dhyana ⁕ Samadhi

The Eightfold Path of Yoga is known as *ashta anga* – literally
'eight limbs'.

Yama is the first stage, the practice of ethical morality, and
is subdivided into five aspects: Ahimsa (compassion), Satya
(truthfulness), Asteya (avoiding theft), Brahmacharya
(continence), Aparigraha (avoiding envy). *Niyama* is the
second limb: the practice of ritual observances, whether it is
prayer, thanks-offerings, temple visits, personal introspection
or meditation. It is once again divided into five disciplines:
Sauca (cleanliness), Samtosa (contentment), Tapas (self-
discipline), Svadhyaya (trying to know thyself and the revealed
scriptures) and Isvarapranidhana (opening oneself to God).

Asana, the third limb, is the most instantly recognisable
aspect of yoga as it is known in the West, the various highly
disciplined physical postures of the body. *Pranayama* is the

control and discipline of breath. *Pratyahara* is the practice of detachment from the external world and the cultivation of a questing introspection. *Dharana* is the ability to concentrate and to cleanse the mind. *Dhyana* is the achievement of an uninterrupted flow of concentration without the distraction of discernible thought processes or objects; just a clean, clear stillness. *Samadhi*, the eighth limb, is the final goal, a sense of union with the world and the divine: the 'peace that passeth all understanding'.

8 LOST BOOKS OF MANI

Living Gospel ❊ Treasure of Life ❊ Pragmateia ❊ Book of Mysteries ❊ Book of Beasts ❊ Epistles ❊ Shaburagan ❊ Ardahang

None of the Books of Mani survive. The sacred texts of Manicheanism (see p.70) were last believed to have been held safe within the Cathars chain of mountaintop fortresses in south-western France, guarded by their ascetic, monk-like Perfects. The books are said to be the true treasure of Mont-ségur destroyed by Catholic Crusaders.

Mani taught by image and sound as well as word: 'I have written books and pictured them in colours. Let him who hears them in words also see them in images, and he who is unable to learn them from writing, let him learn from pictures.' His followers were known as Perfects or Bonshommes, 'Good men', who had no possessions and lived on alms, travelling in order to preach, pray and assist with good works. They would take no pay for any of their labours, and drank no wine or milk, ate no meat, fish, eggs, butter or cheese. They believed that the dark powers ruled this earth and had no truck with priestly authority, relics, icons, churches and had no interest in the Old Testament as a spiritual resource.

They revered the Lord's Prayer, believed in the transmigration of souls and stressed the importance of a good death, both in their daily blessing ('Pray God to make a good Christian of me and bring me to a Good End') and in the practice of a death-bed 'Consolamentum' blessing, which was often followed by a total fast until death.

In this — as in many of their strict vegetarian, non-violent, ascetic beliefs — the Manicheans seem to share many of the practices of the Jain.

OGDOAD

Naunet and Nu ❋ Amunet and Amun ❋
Kauket and Kuk ❋ Hauhet and Huh

The Ogdoad, the eight Egyptian primeval gods of creation, were a curious, inarticulate grouping of four pairs of males and females linked with primordial waters and air and primordial darkness and infinite space. They could be represented as female snakes and male frogs or by eight baboons

The Ogdoad in baboon form salute the rising sun.

raising their forepaws to the rising sun as they made their dawn screech, their prayers to the sun. This symbolism of the eight unknown aspects of creativity would later be embellished by Christian mystics, who developed their own metaphysical Ogdoad: Depth and Silence, Mind and Truth, Word and Life, Man and Church.

8 GREEK WINDS

Boreas (N) ❋ Kaikias (NE) ❋ Eurus (E) ❋
Apeliotes (SE) ❋ Notus (S) ❋ Lips (SW) ❋
Zephyrus (W) ❋ Skiron (NW)

The Tower of Winds in Athens is one of the most enchanting buildings of antiquity – an octagonal 'clocktower' with a wind vane, sundials and a water clock, designed to be seen by the democratic populace in the agora. It later would serve as a bell-tower for a Byzantine church, a lodge for Whirling Dervishes and a study for Lord Byron. The faces of the tower depict the eight winds.

Homer only names the four main winds, but these come down to us full of strong associations. *Notus* (south) wind was associated with violent equinoctial storms and was praised as both the ripener of crops and feared as their potential destroyer. *Zephyrus* (west) wind was associated with the welcome breezes of midsummer, wines and (passionate) love; when Apollo won the love of the handsome Spartan prince Hyacinth, Zephyrus, in jealousy, arranged for a gust of wind to kill the beautiful youth with his own thrown discus. *Eurus*, the unlucky east wind, had associations with death, and so hardly gets a direct mention at all, if the poets can avoid it. *Boreas* (north wind) was associated with winter rainstorms blowing down from the horselands of Thrace and Macedon. It was known as the 'Devouring One' and is sometimes given some sense of leadership of the other winds, either the

Two faces of the Tower of the Winds, depicting Boreas (left) and Skiron.

first-born of the storm-god Aeolus, or siring lesser winds, or at the very least impregnating mares when they backed their hindquarters up against its force.

The four minor, sub-directional winds, the *Anemoi Thuellai*, were also known as the sailing winds. The northeastern *Kaikias* was feared for its dark, violent hailstorms, while the northwestern *Skiron* was known to unleash winter winds and rain. The southeastern *Apeliotes* had more positive associations, as the bringer of the vital rains of spring, and was the preferred sailing wind of that most ancient of all trading nations, the Phoenicians. Southwesterly *Lips* was also thought of as a benign figure, safely escorting the Greek merchant ships to their home ports.

These eight winds were the bedrock of medieval European wind roses (early compasses), lovingly drawn onto the edges of maps and globes and the pavements of piazzas, though their numbers could be expanded to twelve or even sixteen.

THE 8 GREGORIAN CHURCH MODES

Dorian ❊ Hypodorian ❊ Phrygian ❊ Hypophrygian ❊ Lydian ❊ Hypolydian ❊ Mixolydian ❊ Hypomixolydian

The exact origins of this eightfold organisation of modes that completely dominated the church music of medieval Christendom remains contentious. Most authorities accept that the Carolingian court borrowed them from ninth-century Byzantine liturgies, which themselves arose out of the ancient priestly chants of the Near East.

Just as in ancient Greece, generation after generation of writers sought to define their effects on the emotions. *Dorian* was considered to be serious and to tame the passions; *Hypodorian* tended towards the mournful and tearful; *Phrygian* incited passion and led towards mystical reverie; *Hypophrygian* was the mode of tender harmony that tempered anger; *Lydian* was the music of cheerful happiness; *Hypolydian* was the tone of devout and emotional piety; *Mixolydian* united pleasure and sadness; and *Hypomixolydian* aspired to a sense of perfection and secure, contented knowledge.

7

7 ANCIENT VISIBLE PLANETS

Sun ✴ Moon ✴ Venus ✴ Mercury ✴
Mars ✴ Jupiter ✴ Saturn

Our sky-watching, hunter-gathering ancestors had 7 marked out as a number of enormous importance for tens of thousands of years. For this is the number of the visible planets – 'the five wanderers', plus the sun and the moon.

This respect for the 7 became ever more ingrained as the first agricultural civilisations allowed for accurate fixed observations from the calendar-keeping priests, whose temples throughout the ancient Middle East were all equipped with star-watching terraces above their cult chambers. It is an intriguing element within the cult of the 7 that the planets are not all visible at once: Mercury and most especially Venus (whose horns are occasionally visible) are the morning and evening stars. Bright Jupiter, luminous Saturn and the more elusive red Mars belong to full night. So we have always known that we have been watched, influenced and enclosed by these 7 who right from the dawn of our consciousness have intriguingly different characteristics and hours of dominance and passageways through the heavens.

Although most of mankind probably now accepts that the earth is a planet which circles around the sun, and the moon is a planet of the earth, the mystery of our 7 encircling heavens still haunts our imagination. But this once immutable number of 7 keeps changing. First we knocked the seven down to five (as the sun and the moon were taken off the list), then, in relatively modern times, it grew to nine. Uranus was discovered in 1781, followed by Neptune in 1846, then Pluto in 1930 (though this was later demoted to a dwarf planet to bring us back down to eight planets). So, currently, we have eight planets and five dwarf planets (Ceres, Pluto, Haumea, Makemake and Eris), as well as five named moons orbiting around Pluto.

The seven ancient planets are identified in this medieval engraving with the days of the week: Saturn (Saturday), Jupiter (Thursday), Mars (Tuesday), Sun (Sunday), Venus (Friday), Mercury (Wednesday), Luna (Monday).

7 NOTES

Do ❀ Re ❀ Mi ❀ Fa ❀ So ❀ La ❀ Ti

As early as the seventeenth century, European musicians believed that this mnemonic for teaching musical pitch was derived from a Muslim source, though we now think this may itself lead back to a Sanskrit Bronze Age hymn. There is an equally strong tradition that it came from the first letters from each phrase of an eighth-century hymn to Saint John which goes: 'So that these your servants can, with all their voice, sing your wonderful feats, clean the blemish of our spotted lips, O Saint John' – or, rather, in Latin, '*Ut queant laxis resonare fibris, Mira gestorum famuli tuorum, Solve polluti labii reatum, Sancte Ioannes*'.

However, for most of us the whole seven-note mnemonic is intrinsically bound up in Julie Andrews' teaching the Von Trapp children to sing in the film *The Sound of Music*. This is one of the most beloved propaganda films of all time, creating an emotional case for excluding the inhabitants of the beautiful mountain scenery of the Austrian Alps from any complicity with the war crimes of Nazi Germany. 'Doe a deer, a female deer, Ray a drop of golden sun, etc.'

FATE OF THE 7 DAYS

Monday's child is fair of face ❀ Tuesday's child is full of grace ❀ Wednesday's child is full of woe ❀ Thursday's child has far to go ❀ Friday's child is loving and giving ❀ Saturday's child works hard for a living ❀ And the child that is born on the Sabbath day is bonny and blithe, good and gay

This poem was first printed in a collection of Devon folk-tales in 1838, though it had a widespread English oral tradition for many centuries before this. Should you recite it to a child who turns out to be born on a Wednesday, you should know that there is a useful version that swaps fates with Friday's child. You can also, at will, exchange the Scots-sounding phrase of 'bonny and blithe' for 'happy and wise'.

SHAKESPEARE'S 7 AGES OF MAN

Infant ✿ Schoolboy ✿ Lover ✿ Soldier ✿
Justice ✿ Pantaloon ✿ Second Childhood

His acts being seven ages. At first, the infant,
Mewling and puking in the nurse's arms.
And then the whining school-boy, with his satchel,
And shining morning face, creeping like snail
Unwillingly to school. And then the lover,
Sighing like a furnace, and with a woeful ballad
Made to his mistress' eyebrow. Then a soldier,
Full of strange oaths, And bearded like the pard,
Jealous in honour, sudden and quick in quarrel,
Seeking the bubble reputation
Even in the cannon's mouth. And then the justice,
In fair round belly with good capon lined,
With eyes severe and beard of formal cut,
Full of wise saws and modern instances:
And so he plays his part. The sixth age shifts
Into the lean and slipper'd pantaloon,
With spectacles on nose, and pouch on side;
His youthful hose, well saved, a world too wide
For his shrunk shank; and his big manly voice,
Turning again toward childish treble, pipes

And whistles in his sound. Last scene of all,

That ends this strange eventful history,

Is second childishness and mere oblivion,

Sans teeth, sans eyes, sans taste, sans everything.

This speech is often imagined to be Shakespeare's melancholic benediction from his death bed, some time after he had made out that vindictive will that left all to his children and his sister but just 'his second best bed to his wife'. It was, however, penned much earlier in his life, for Jaques to speak in that oft-quoted 'All the world's a stage' speech: Act II, Scene vii of *As You Like It*.

The Seven Ages of Man depicted in a French woodcut, c. 1510, by Bartholomeus Anglicus, author of the first encyclopedia.

7 ANGELS OF PUNISHMENT FROM THE TESTAMENT OF SOLOMON

Kushiel (Rigid One of God) ※ Lahatiel (Flaming One of God) ※ Shoftiel (Judge of God) ※ Makatiel (Plague of God) ※ Hutriel (Rod of God) ※ Puriel (Fire of God) ※ Rogziel (Wrath of God)

The *Testament of Solomon* is an intriguing work, not listed in either the Christian or the Hebrew canon of Holy Books, but probably written down at about the time of Christ, in that hybrid Judaeo-Hellenistic world created in Levantine cities such as Antioch and Alexandria. Whatever its origins, its influence did leak into the popular imagination of the Middle East as the *Testament of Solomon* provided wonderfully intriguing details about King Solomon's magical court, where demons and djinn were harnessed to complete the building of the Temple at Jerusalem.

7 ARCHANGELS

Michael ※ Gabriel ※ Raphael ※ Chamael ※ Jophiel ※ Zadakiel ※ Izrael

Michael, 'who is like God', is also referred to as the Prince of Seraphim, and usually depicted sword in hand, sometimes as a Roman soldier. His feast day is 29 September. *Gabriel*, 'strength of God', is often depicted holding a lily. He is, of course, the archangel of the Annunciation to the Virgin Mary and the angel who also communicated the Koran to the Prophet Muhammad. His Christian feast day is 24 March. *Raphael*, 'medicine of God', is often depicted with a staff (decorated with a red ribbon) and in the garb of a pilgrim, sometimes with Santiago-fashion cockle sHells sewn into his cloak. His Christian feast day on 24 October.

The other four archangels are less well-known, and *Chamael*, *Jophiel*, *Zadakiel* and *Izrael* are sometimes listed as Uriel, Raguel, Saraquel and Jeramiel.

Jewish scholars believed that knowledge of the archangels' names, and their number, was passed on to them by star-gazing Chaldean high priests during the Jews' exile in Babylon. The Christian Book of Revelations seems to follow this tradition when Saint John writes of the 'Seven Angels who stood before God' (8:2). This seems to give these seven archangels a prime role in the hierarchy of Heaven, though later writers mark the archangels well down the ranks, in the second choir of the third hierarchy of nine angel types (*see 9*).

Muslim beliefs do not include such an emphatic ordering of the angelic hierarchy, in part because there are so many of them – two recording angels for each person, and one before and one behind every believer, as attested in Sura 13:11 of the Koran. But Islamic, like Jewish and Christian, tradition has names for the seven principal archangels.

Jibral (Gabriel) ❋ Mikail (Michael) ❋ Israfil (Raphael) ❋
Azrail ❋ Ridwan ❋ Maalik ❋ Sariel

There is some hierarchy among these with a trio of principals. *Jibral/Jibril* (Gabriel) gets top billing, as one would expect for the announcer of revelations (most especially the Koran), while *Mikail* (Michael) tends to be less martial than the Christian view of him and much more of a helper in times of hunger and distress. *Azrail/Izrail/Azrael*, also known as the Malak al'Mail, is the angel of death, responsible for parting the soul from the body.

The remaining four archangels are *Israfil* (Raphael), who will blow the last trump at the end of time; *Ridwan*, the guardian of the gates of Heaven; *Maalik*, guardian of the gates of Hell (Jahannam); and *Sariel*, an angel of death who records one's name at birth and erases it when you die.

7 CHEMICALS OF THE ALCHEMIST'S ARCANA

Sulphuric Acid ❋ Iron oxide ❋ Sodium carbonate ❋
Sodium nitrate ❋ Liquor hepatis ❋ Red pulvis solaris ❋
Black pulvis solaris

The alchemists' vocabulary does not always translate directly into a modern formula. They were keen on *Natron*, which was a generic word that included both the salts of sodium carbonate and sodium nitrate. *Vitriol*, however, is what we know as sulphuric acid and *Aqua Fortis* was nitric acid. *Black pulvis solaris* was formed from ground black antimony (stibnite – a sulphide of antimony) mixed with ground sulphur. *Red pulvis solaris* was a mixture of mercury (which could be extracted by heating cinnabar) and sulphur.

The alchemists also made a strong connection between the seven prime metals and the planets. The sun was linked to gold, the moon to silver, Mars to iron, Mercury to quicksilver, Saturn to lead, Jupiter to tin and Venus to copper.

ORDERING THE 7 HEAVENS

Moon ❋ Mercury ❋ Venus ❋ Sun ❋
Mars ❋ Jupiter ❋ Saturn

The ordering of the seven heavens is one of the mysteries of each culture, especially as it appears to be linked to everything. The Chaldeans created a very influential list, ordering the moon, Mercury, Venus, the sun, Mars, Jupiter, Saturn. This seems to reflect a very exact ascending order, based on the observed length of time that they circled the earth. The *moon*, as we know from our months, is 29.5 days, *Mercury* 88, *Venus* 224.7, the *sun* (the length of our year) is 365.25, *Mars* 687.1, while *Jupiter* is 12 years and *Saturn* 29.5.

This remains virtually the pattern we follow today, apart from the reordering of the sun-day as the first not the fourth. This ordering seems to have been achieved in the Hellenistic East, where we know that the Astrologers of Alexandria had created a hierarchy of sun, moon, Mars, Mercury, Jupiter, Venus, Saturn, which ascribed an order of dominant deities to each of the progressive hours of the daylight.

7 CIRCUITS OF THE KAABA

The most important of the ritual actions of a Hajj pilgrim to Mecca is to make the seven circuits of the Kaaba, a four-square shrine that is currently draped in black cloth embroidered with Koranic calligraphic verse. The Prophet Muhammad made it his habit to kiss the Black Stone (built into a corner of the shrine) as he passed each time, though

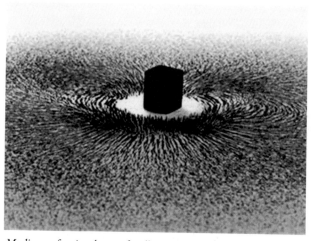

Muslims performing the tawaf, walking seven circuits of the Kaaba, which in this photo seems to exercise an almost gravitational pull.

due to the vast pressure of modern crowds this is no longer possible for most pilgrims.

The origin of the Black Stone is shrouded in mystery, though pious Muslims believe it was sent from Heaven to be an altar for Adam and Eve, then was rediscovered by Abraham (assisted by an angel), who instructed his son Ishmael (the patriarch of all the Arab tribes) to set it up again as an altar. It remained venerated during the many centuries of pagan worship – one of many such sacred stones of ancient Arabia – before being brought into the fold of Islamic worship by the Prophet. The second Caliph, Omar, used to formally address it: 'No doubt, I know that you are a stone and can neither harm anyone nor benefit anyone. Had I not seen God's Messenger kissing you, I would not have kissed you.'

7 CLASSICAL HEROES WHO VISITED THE UNDERWORLD

Aeneas ❖ Odysseus ❖ Orpheus ❖ Dionysus ❖
Heracles ❖ Psyche ❖ Theseus

Virgil has *Aeneas* descend with the sibyl into the underworld at the sulphur-ridden crater of Avernus near Cumae to speak to his dead father. *Odysseus* makes it as far as the banks of the River Charon. However, *Orpheus* succeeds in charming Pluto and Persephone with his music and almost succeeds in extracting his lover Eurydice from the gates of Hell but on his return to the light gives birth to a mystery religion complete with a transformational initiation rite, hymns and a priesthood who remain poor outcast wanderers, renouncing their taste for meat and women.

Dionysus's descent feels like an earlier episode in this same half-understood Orphic religion, though dance replaces music and Dionysus is successful in rescuing his mother Semele and placing her in the heavens. *Hercules* is in Hades on a mission to steal the hound of Hell (Cerberus), but again seems to follow in the spiritual footsteps of Orpheus by descending into the underworld via Eleusis and its mystery cult.

Looking beyond the Aegean, and this list of seven, are the much older stories of Gilgamesh's journey to Hell and the Sumerian-Babylonian goddess Inanna's descent.

7 CHURCHES OF SAINT JOHN

Ephesus ❊ Smyrna ❊ Thyatira/Akhisar ❊ Pergamon/
Bergama ❊ Sardis ❊ Philadelphia ❊ Laodicea

The Seven Churches (communities of Christians) are all in Asia Minor (modern Turkey) to whom John addressed his Revelations: 'Write in a book what you see and send it to the seven churches, to Ephesus, to Smyrna, to Pergamum, to Thyatira, to Sardis, to Philadelphia, and to Laodicea.'

7 WONDERS OF THE ANCIENT WORLD

The Pharos of Alexandria ❊ The Colossus of Rhodes ❊
The Mausoleum of Halicarnassus ❊ Temple of Artemis at
Ephesus ❊ The Statue of Zeus at Olympia ❊ The Hanging
Gardens of Babylon ❊ The Great Pyramid of Giza

You can trace the foundations of the *Temple of Artemis* at Ephesus but this is one of the least rewarding things at this dazzling site; likewise, you can look at the foundation pit of the *Mausoleum of Halicarnassus* at modern-day Bodrum, though the seafront castle which quarried the tomb is much more fun. The shape and even the exact location of the *Colossus of Rhodes* (a brazen statue of Helios), and the *Hanging Gardens of Babylon* remain unknown. The lower storey of the *Alexandria*

Pharos (lighthouse) is visible, as it became encased in a fortress, but there is nothing to quicken the soul. Phidias's *Statue of Zeus* survives only on coins. So the only wondrous ancient wonder left to the world is the *Great Pyramid of Giza*.

The Seven Wonders of the Ancient World – a little artistic licence has to be involved, as only the Great Pyramid of Giza survives.

The earliest literary references to the seven wonders comes from Herodotus (the half-Carian, half-Greek historian of the Greek resistance to the Persian invasion, who was well travelled enough to have made some valid comparisons) and Callimachus (a Hellenistic scholar-poet-librarian). Though these two might have allowed local feelings to sway their final judgement ... for Callimachus was from Alexandria and Herodotus was born in Halicarnassus.

7 COUNCILS OF THE CHURCH

First Council of Nicaea (325) ❉ First Council of Constantinople (381) ❉ Council at Ephesus (381) ❉ Council at Chalcedon (451) ❉ Second Council at Constantinople (451) ❉ Third Council at Constantinople (680–81) ❉ Second Council of Nicaea (787)

These seven Church Councils are recognised by the Catholic, Orthodox and Anglican churches. Indeed, Pope Gregory asserted the collective authority of the first four Church Councils: 'I confess that I accept and venerate the four councils in the same way as I do the four books of the Holy Gospel.'

The *First Council of Nicaea* was assembled at the invitation of the Emperor Constantine and consisted of either 250, 270, 300 or 318 fathers of the church. Its greatest and most enduring achievement was the Nicene Creed. The *First Council of Constantinople* met at the invitation of the Emperors Gratian and Theodosius I and revised this Creed. The *Council at Ephesus* was summoned by the co-Emperors Theodosius II and Valentinian III and proclaimed the Virgin Mary as Theotokos: 'one who gave birth to God'.

The *Council at Chalcedon* (just across the water from Constantinople on the Asian shore of the Bosphorus) tried to define the two natures of Christ, human and divine, and adopted the Chalcedonian Creed. The *Second Council of Constantinople* condemned the Nestorian and Monophysite definitions of the two natures of Christ as heresies. The *Third Council of Constantinople* repudiated Monothelitism (an attempt at a compromise over the two natures of Christ) and affirmed that Christ had both human and divine wills. The *Second Council of Nicaea* restored the veneration of icons and brought to an end the first iconoclastic period. (It has been subsequently rejected by some of the Protestant communions who prefer to list the Council of Constantinople of 754 – which condemned the veneration of icons.)

---

7 DAYS OF THE WEEK

Monday/Lundi ❋ Tuesday/Mardi ❋ Wednesday/Mercredi ❋ Thursday/Jeudi ❋ Friday/Vendredi ❋ Saturday/Samedi ❋ Sunday/Dimanche

Our seven-day week is a straight inheritance from very ancient Babylonian and Jewish traditions that took the seven planets as one of the ordering principles of humanity and divinity. The main alternatives were the Egyptian ten-day week, the Germano-Celtic nine-night week and the eight-day week of the Etruscans. The latter was inherited by the Romans, for it allowed for a specific market-day, which enabled country-dwellers to come to the cities and sell fruit and vegetables (which lasted only eight days). During Julius Caesar's calendar reforms the seven-day week was introduced from the Near East, though it ran alongside the old Etruscan traditions until the time of Constantine.

And some time during that period, between 200 and 600 AD, the current charming muddle of English names was hatched out, part honouring the Roman pantheon and part the Norse-German deities. For *Monday* is moon day, *Tuesday* is the day of Tiw/Tyr's day (the heroic Tuetonic sky god),

Wednesday is Woden/Odin's day (the Teutonic/Norse god of knowledge and war), *Thursday* is the day of Thor (the Teutonic smith-god of thunder), *Friday* is the day of Frija/Freyr (the Teutonic goddess of fertility), *Saturday* is Saturn (the father of Zeus)'s day, and Sunday is of course the sun's day.

The same process happened in France, ossifying that peculiar junction point between Roman paganism and the new Christian order. So the French have *Lundi* (from the Latin *dies Lunae*, or moon day), *Mardi* (*dies Martis*, or Mars day), *Mercredi* (*dies Mercurii*, or Mercury day), *Jeudi* (*dies Jovis*, or Jupiter day), *Vendredi* (*dies Veneris*, Venus day), *Samedi* (*dies Saturni*, Saturn day) and *Dimanche* (*dies Dominicus*, day of the Lord).

In the well-ordered Christian state of Byzantium, all these pagan relics were ditched in favour of days 1, 2, 3 and 4, followed by Paraskene (preparation), Sabbaton and finally Kyriaki (God's day). These remain the days in modern Greek.

7 DEADLY SINS OF CHRISTENDOM

Gluttony ❖ Pride ❖ Greed ❖ Lust ❖ Envy ❖ Anger ❖ Sloth

The Seven Deadly Sins could collectively be represented by the biblical Leviathan, whose origin looks back to the Canaanite terror of the deep – the seven-headed serpent Lotan destroyed by the great god Baal. In medieval imagery, *Lust* was represented by an ape, though this animal could also express idolatry and, when given an apple, the expulsion from paradise. An ass playing a lyre was used by Romanesque sculptors to represent *Pride*. A bear could be used to represent either *Gluttony*, *Lust* or *Anger,* while by reverse logic a bee could represent *Sloth*. The boar could also symbolize *Lust*.

List-making is an ancient art and scholars have traced the seven deadly sins as moral manifestations of the seven evil spirits, first codified by King Solomon in his proverbs, then

*The Seven Deadly Sins envisioned by the Flemish painter
Peter Brueghel the Elder. This one is merely Avarice.*

reworked by Saint Paul in his rather stern letter to the Galatians. A hermit monk, one Evagrius Ponticus, turned them into eight spiritual temptations that might beset an ascetic (a bit like the demons that tormented Saint Anthony). But it was Pope Gregory the Great in the sixth century who must be credited with the edition that survives today, as well as the seven positive virtues – *Faith, Hope, Charity, Fortitude, Justice, Prudence, Temperance* – and the seven defences:

Abstinence against Gluttony ❖ Humility against Pride ❖
Liberality against Greed ❖ Chastity against Lust ❖
Kindness against Envy ❖ Patience against Anger ❖
Diligence against Sloth

7 DESTRUCTIVE SINS OF ISLAM

Worship other gods along with Allah ✳ Practice sorcery
✳ Kill the life which Allah has forbidden except for a just
cause ✳ Eat up with usury ✳ Eat up an orphan's wealth ✳
Treason and flight from the battlefield ✳
False accusation against a chaste woman

This sorts of list is part of collective Islamic tradition, often
created as a negative notice-board in response to the Five
Pillars of Islam and the Seven Deadly Sins so beloved by
Christian medieval scholarship. There are many variants but
all include usury (*riba*), murder and the sin of *shirk* (associat-
ing others with Allah).

7 COLOURS OF THE VISIBLE SPECTRUM

Red ✳ Orange ✳ Yellow ✳ Green ✳ Blue ✳ Indigo ✳ Violet

These seven colours can be remembered through the mem-
nonic 'Richard Of York Gave Battle In Vain'.

SNOW WHITE'S 7 DWARFS

Bashful ✳ Doc ✳ Dopey ✳ Grumpy ✳
Happy ✳ Sleepy ✳ Sneezy

The Brothers Grimm who first recorded this old story
never listed the names of their seven dwarfs, so we all remain
happily free to invent and elaborate. The 1937 film ver-
sion gave them the names Bashful, Doc, Dopey, Grumpy,
Happy, Sleepy and Sneezy; that of 1912 produced the more
Germanic list of Blick, Flick, Glick, Plick, Quee, Snick and
Whick; while the 1965 remake had Axelnod, Bartholomew,
Cornelius, Dexter, Eustace, Ferdinand and George.

7 HILLS OF ROME

Aventine ❋ Caelian ❋ Capitoline ❋ Esquiline ❋
Palatine ❋ Quirinal ❋ Viminal

Rome was founded as a network of seven villages perched on seven hills, so that Sabines, Latins and Etruscans could all benefit from the markets usefully arranged in the low-lying land in between them. The *Palatine* was the central hillock, the *Capitoline* overlooked the marshy field of Mars and the *Aventine* was hard against the banks of the River Tibur. In all honesty the other four hills are not so distinct, just a series of interlinked spurs, but it has always been immensely propitious to have a unit of 7 in your foundation myth, like the very first civilisation born in Mesopotamia, Sumeria. Rome doubled up by honouring a list of its first seven kings, beginning with Romulus (753–716), Numa Pompilius (715–674), Tullus Hostilius (673–642), Ancus Marcius, Lucius Tarquinius Priscus and Servius Tullius and finishing with Tarquinius Superbus.

The power of Rome further spread the allure of a city being founded on seven hills, so that most of the great cities of the world – say, Moscow, Lisbon, Jerusalem, Istanbul or Barcelona – have a story of seven hills. Others which have hardly a hill at all, like Mumbai, are said to be founded on seven islands.

THE 7 HAUSA CITIES

Daura ❋ Zaria ❋ Biram ❋ Kano ❋
Katsina ❋ Rano ❋ Gobir

These are the seven city states of the Hausa people of Central West Africa whose historic territory extends across Nigeria, Niger and several other modern nations.

7 VOWELS

Alpha ✷ Epsilon ✷ Eta ✷ Iota ✷
Omicron ✷ Upsilon ✷ Omega

The vowels have always been linked with the seven heavens, most famously in Hebrew, where the seven unwritten vowels created the sound for God – Jehovah. The link between the language of man and the presumed languages of the seven Heavenly spheres has always been speculated upon. However, it is one of the more arcane secrets of the mystics which of the seven planets is linked to which vowel.

7 GATES OF HELL THROUGH WHICH INANNA SURRENDERS HER DIVINE REGALIA

Crown ✷ Necklace of lapis lazuli ✷ Breast beads ✷
Breastplate ✷ Gold ring ✷ Sceptre-rod ✷ *Pala* robe

The sensual goddess of love, lust and power in the ancient Middle East was known by many titles, such as Ishtar and Astarte, though one of her most ancient names was Inanna, derived from the Sumerian *Nin-Anna* 'Queen of Heaven'. She was always strongly associated with the planet Venus and for centuries the first dynasties of kings only claimed their authority to rule by joining her in her sacred bed on the tenth day of the spring New Year festival. She had many lovers and was known to enslave them all.

One of her most famous, but most capricious, adventures was to journey down to the Underworld. At each gateway that guarded the road to Hell, she was gradually stripped of her regalia of power, her complaints being answered: 'Be Quiet, the ways of the underworld are perfect, they may not be questioned.' So she meekly surrendered first her crown, then her lapis lazuli necklace, then her double strand of beads about her breast, then her breastplate, then her gold ring of power, then her sceptre of dominion (a lapus lazuli measuring rod) before being finally stripped of her *pala*, the mantle of Heaven. Finally, naked and powerless, she begs for mercy from her sister, and the seven judges, but is condemned to be impaled (or hung upon a hook), whereupon at that instance the lustful frenzies that empower our world are silenced.

7 SENSES

Smell ❋ Hearing ❋ Sight ❋ Taste ❋
Touch ❋ Speech ❋ Animation

The last two are sometimes replaced by 'intuition' and 'balance' or occasionally left out and the first five listed as the natural wits of mankind. 'All our knowledge begins with the senses, proceeds to the understanding and ends with reason', pronounced Immanuel Kant.

7 GRADES OF MITHRAIC INITIATION

Corvus – Raven – Mercury ❋ Nymphus – Bride – Venus ❋
Mile – Soldier – Mars ❋ Leo – Lion – Jupiter ❋
Perses – Persian – Moon ❋ Heliodromus – Runner of the
Sun – Sun ❋ Pater – Father – Saturn

The Mithraic cult is a peculiar spin-off from the normal structures of world mythology. For Mithras (literally 'the friend') was a young hero-god who came out of Iran but was not ordained to fall as a sacrifice like Jesus, Adonis, Attis or Tammuz. Rather, he was a martial warrior, born from out of a rock, who succeeded in sacrificing his opponent, the bull who is cornered in a cave lit by a ray of sun. The spilling of bull's blood (the tauroctony sacrifice which is believed to have drenched an initiate, who crouched in a pit below the sacrificial altar, in blood) seems to make no particular sense to the cults of the Persian Empire, whose religion was famous for its bloodless fire temples and somewhat obsessed with ritual purity. Possibly one should look beyond Iran for the origins of Mithras, towards the traditional practices of the nomadic steppe-lands where vast sacrifices of animals to the sun and the heroic ancestor took place.

Mithraic temples have been found throughout the Roman Empire (one rather sad Mithraic chapel has been exposed in the City of London) and were popular amongst the officers and soldiers of the Legions, though bizarrely the hero Mithras is depicted as a Persian (the Persian Empire was Rome's only real military adversary, in both its Parthian and Sassanid manifestations). Mithras always wears the trousers worn by a Persian knight and is usually sacrificing a cosmic bull, his left knee on the bull's back, his hand in the bull's nostrils and a stabling knife about to pierce the bull's neck through the

Mithras – once a Persian, always a Persian.

shoulder. He is often escorted by Cantes and Cantopates, the torch-bearers of life and death, and in some regions (such as Thrace) a Heavenly knight-like Mithras hero is helped in his hunt of the bull by a dog and a snake, while sometimes a scorpion attacks the bull's testicles.

7 PLEIADES AND HYADES

Maia ✳ Electra ✳ Celaeno ✳ Taygeta ✳ Maia ✳
Merope ✳ Asterope ✳ Alcyone

The Pleiades – daughters of the giant Atlas and the sea nymph Pleione – took their lives in despair at the fate of their other seven sisters (the Hyades) and their father's eternal imprisonment as one of the foundation blocks of Heaven. Raised to Heaven by the gods and then ravished by Zeus and his brother Poseidon, they gave birth to many famous heroes and gods. *Maia*, for instance, was considered the mother of Hermes. In the ancient calendar of star-watching they are the most obvious cluster of all (the star of stars to the Babylonians), a tight saucepan in shape within the constellation of Taurus. They heralded autumn (in the northern hemisphere), warning men to beach their boats and leave the seven seas unsailed over the winter. This may have been the origin for their name, as in this verse from Hesiod: 'And if longing seizes you for sailing the stormy seas, when the Pleiades flee mighty Orion, and plunge into the misty deep, and all the gusty winds are raging, then do not keep your ship on the wine-dark sea, but – as I bid you – remember to work the land.'

Aesyle ✳ Cleeia ✳ Eudora ✳ Pedile ✳
Phaeote ✳ Phyto ✳ Polyxo

The Hyades (above), the sisters of the Pleiades, provide the other bright group within the constellation of Taurus, a

V-shaped group roughly where the eyes of the great Bull might be imagined to be. Normally a cluster of seven, they can also be named as a group of five and sometimes stretched to number fifteen. Their heliacal 'rising' coincided with the rainy seasons of March and November, which may be why they were connected with the fertility cult of Liber Pater/ Dionysus/Bacchus. The Romans affectionately names them 'the little pigs' and the Arabs also knew them as a benign influence, Al-Kilas – 'the little she-camels'.

THE 7 LIBERAL ARTS

Grammar ※ Rhetoric ※ Logic ※ Arithmetic ※
Geometry ※ Music ※ Astronomy/Cosmology

The Seven Liberal Arts divide between the trivium of *logic, rhetoric* and *grammar* and the quadrivium of *arithmetic, music, geometry* and *cosmology*. The Trivium were the arts considered necessary for the creation of an active citizen of the ancient world, well educated enough to be able to analyse what was being said, check it for rationality and to be able to speak and answer in his turn. With the addition of the quadrivium by the scholastics of the early medieaval age, the whole basic course structure and purpose of university education was established – which was to create an aware citizen. The system endured, more or less unchanged, right through to nineteenth-century Europe.

7 ANCIENT THINGS FOR NOWRUZ

Sharab ※ Shir ※ Sharbat ※ Shamshir ※
Shemshad ※ Sha ※ Shaddaneh

In the Persian Empire, the chief priest of the Zoroastrians would start the spring Nowruz celebrations by ceremoniously presenting the ruler with seven symbolic gifts on the first day that light equals darkness: a golden goblet of wine, a golden bowl of milk, a golden ring, a purse of golden coins, a sword, a bow and a sacred bowl of sprouting wheat. In more domestic celebrations, the Nowruz table would welcome spring with wine, milk, sherbet (a sweetened fruit juice), a sword, a bow, a candle and sprouting green seeds arranged on a house altar.

To the Zoroastrians, these can represent many things: the seven elements of life: fire, earth, water, air, plants, animals and humans. As well as signifying life, health, wealth, abundance, love, patience and purity, they also have correlations to the seven ancient planets: Mercury, Venus, Mars, Jupiter, Saturn, the sun and moon.

7 SAGES OF ANCIENT GREECE

Thales of Miletus ❊ Bias of Priene ❊ Heraclitus of Ephesus ❊ Cleobolus of Lindos ❊ Solon of Athens ❊ Pittacus of Mytilene ❊ Periander of Corinth

This is an acceptable list, though there are many variants, not least because the great kings of antiquity liked to keep seven sages – in Greek, *hepta sophoi*, in Latin, *septem sapientes* – around their courts.

There also seem to have been competitions for sage advice in verse, which allowed various pantheons of seven sages to be formed. This was especially true of the Pythian Games held in honour of Apollo, the god of wisdom. Some of the most pithy couplets were then carved on the porch of the Temple of Apollo at Delphi. The two best known, as reported by that great guidebook writer Pausanias, are 'Know thyself' and 'Nothing in excess'.

7 SACRAMENTS OF THE CATHOLIC AND ORTHODOX CHURCHES

Baptism ❊ Confirmation ❊ Eucharist/Communion
❊ Holy Orders ❊ Confessional penance ❊ Marriage ❊
Extreme unction (anointing of the sick and dying)

The authority of both the Catholic and Orthodox churches is united in recognising the especial nature of the seven great sacraments, while not denying that other actions of the church may also have a sacramental nature. It is also one of the frontiers of the Reformation, for all Protestant churches are united in believing that the two sacraments witnessed by the Gospels (Baptism and the Eucharist/Communion) must be given priority, if not total dominance in the rites. If you are ever in Edinburgh, do not miss the opportunity to look at Poussin's series of Sacraments in the Scottish National Gallery, opposite the train station.

Ordination – from Poussin's series of Sacraments.

7 SEAS

North Atlantic ❈ South Atlantic ❈ Arctic ❈ Antarctic ❈
Indian Ocean ❈ North Pacific ❈ South Pacific

These vast Oceans are the seven seas that we now list –however, the concept of the seven seas is ancient and also very variable. We know the Sumerians had a list (from a reference in a hymn of the Enheduanna) but not what was on it. By the time of the Phoenicians, there was a canonical list for the seven seas within the Mediterranean, upon which their black ships traded. Working west from their homeland, there was the Aegean, Ionian and Adriatic whilst west of Sicily stretched the Tyrrhenian, Ligurian, Balearic and sea of Alboran (the straits of Gibraltar).

For a Muslim Arab trader the seven seas referred to that vital sinew of trade that took them east to the coast of China, beginning with the Persian Gulf, then the Gulf of Khambhat (Sind and Gujerat), Harkand (the Indian Ocean after the Bay of Bengal), Kalah (the Malacca straits), Salahit (the straits of Singapore), Kardani (the waters of Siam) and Sanji (the South China sea). Medieval Christian traders, such as the Venetians and Genoese, made lists of seven that included the Adriatic, Black Sea, Caspian Sea, Red Sea, Mediterranean, Arabian Gulf and Indian Ocean.

7 TRIALS OF TRAVEL

Don't ever go a-roving, O my friend, if you'd
Escape the seven circles of the traveller's Hell.
The first's a haunt of homesick thoughts and solitude,
The second's where your fears for far-off family dwell.
The third's a den of thieves, the fourth's the latitude
Where rogues rip off an unsuspecting clientele.

> Then comes the Hells of lonely nights and nasty food.
> The last, and worst: the Hades of a bad hotel.

This familiar litany was recorded by Tim Mackintosh-Smith in a souk in Mauritania.

SAPTAPADI: 7 STEPS OF A HINDU WEDDING

Take care of each other ❄ Grow together in physical joy and mental strength ❄ Jointly protect, nurture and grow resources ❄ Share both joy and sorrow together ❄ Take mutual delight in children and shared duty to each other's family ❄ Live together for ever ❄ Strive to be truthful and honest so that friendship will grow beside love and knowledge of each other

The traditional Hindu wedding rite is performed around a sacred fire, where the groom, tied to the bride, leads them in a circuit around the fire altar, each step coinciding with another in the series of seven promises. These can be seen as both commitments and blessings that are offered up to the new couple by the priest and the surrounding congregation. As each blessing is finished, the couple throw puffed rice or corn into the fire as a sacred symbol of their agreement.

7 VALLEYS OF THE BAHAI

Search ❄ Love ❄ Knowledge ❄ Unity ❄ Contentment ❄ Wonderment ❄ Extreme poverty and Absolute Nothingness

The Seven Valleys takes a spiritual pilgrim on a journey. They are set out in a mystical treatise composed in Baghdad in 1860 by Baha Ullah, framed in response to the questions of a Sufi sheikh who was also an orthodox Muslim judge.

Both would have immediately understood the links to the seven Sufi stages towards the perfection of a soul: compulsion, conscience, inspiration, tranquillity, submission, servant, perfection. *The Seven Valleys* is full of such spiritual and poetic allusions and Koranic references, and is the foundation document of the Bahai faith.

7 SUPREME WORKS OF SHAKESPEARE

Henry IV ❊ Hamlet ❊ Measure for Measure ❊ Othello ❊ King Lear ❊ Macbeth ❊ Antony and Cleopatra

'Just as there are seven wonders of the world and seven deadly sins, so there are (in my opinion) seven supreme peaks achieved by Shakespeare,' wrote Giuseppe de Lampedusa, author of *The Leopard*. He also added that, 'If I was told that all the works of Shakespeare had to perish except one that I could select, I would first try to kill the monster who had made the suggestion; if I failed, I would then try and kill myself: and if I could not manage even this, well then I would choose *Measure for Measure*.'

6

HEXAGONS AND IDEALS

Six appears enduringly solid as a metaphor: the arms fully outstretched to mark out a fathom of rope (6 feet); the faces on a dice; the walls of a honeycomb, expanded in a lattice-like grid for all beehives; and confirmed by a pair of triangles locked together to form a rigid hexagon. Therefore 6 tends to be associated with a sense of space-occupying materiality. The Pythagoreans loved to point out that 6 is uniquely the sum of the first three numbers $(1 + 2 + 3)$ and also the multiple $(1 \times 2 \times 3)$. It is also, as Euclid established, the smallest perfect number.

However, for all these virtues, the 6 can appear curiously empty in the mythic imagination of mankind. It is almost if that trick of hands, during a bored moment in a pub, where you kiss six coins together in a circle and so create the perfect void of a seventh in the centre, hints that one is never quite happy enough with just a 6, so even with something as elemental as the Six Days of Creation, one knows that this is not quite complete in itself but will lead to a seventh day of rest before the cycle is free to continue.

6-POINTED STAR OF DAVID

The six-pointed star – a hexagram – has been a symbol of Judaism since the seventeenth century. Its decorative use, on synagogues and tombstones, goes back far earlier, however, with examples dated to the fourth century AD, and the symbol also has early traditions in Hinduism, Buddhism and Jainism. It was (and is) a popular Occult motif, too, and is a dominant decorative feature in Masonic temples, through its association with the Temple of Solomon. Confusingly, both the hexagram and pentagram are known as Solomon's Seal.

The meaning of the six-pointed star is oddly obscure, even in its Jewish tradition. Its association with David, as the Shield of David (a title for God), seems to have emerged in medieval Kabbalistic literature. Its symbolic associations include the arrangement of the Seder plate at Passover; the number 7 (the star forms seven spaces) and the Menorah; and the number 12 (its nodes) representing the tribes of Israel.

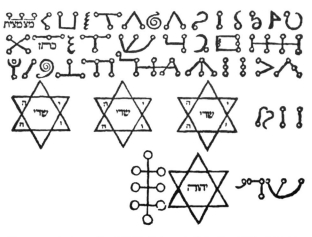

Hexagrams in the medieval Kabbalistic work, Sefer Raziel Ha-Malakh.

The political adoption of the Star of David as a symbol of Judaism emerged in Vienna during the seventeenth century, when it seems to have displaced the pentagram. Its use gained traction in nineteenth-century France, where it was chosen by Zionists and later adopted internationally. Despite, and in defiance of, its Holocaust associations – Nazi Germany forced Jews to wear the symbol for identification – it was adopted by Israel as its flag on the state's foundation in 1948.

6 DYNASTIES

Wu ❋ Jin ❋ Liu Song ❋ Qi ❋ Liang ❋ Chen

The Six Dynasties Period can be compared to Europe's Dark Ages as it stands between the two great cultural blocks of the Han (contemporary to the Roman Empire) and the Tang (about the time the Islamic Empire emerged). The first half of this period is sometimes known as the Three Kingdoms Period and the dates can vary according to eastern and western divisions between the states, but, give or take a year or two, the *Wu* ruled from 22–285, the *Jin* from 265 to 420, *Liu Song* from 420 to 479, the *Qi* from 479 to 502, *Liang* 502 to 557 and the *Chen* from 557 to 589.

6 CONFUCIAN CLASSICS

Book of Changes ❋ Book of Documents ❋ Book of Poetry ❋ Records of the Rites ❋ Spring and Autumn Annals ❋ Records of Music (missing)

As the son of an officer in the service of his ducal state, Confucius's life was informed by the middle-class respect for textual learning. Even in his youth (he was born in 551 BC), he yearned for a golden past of decency, harmony and respect, and dressed in eccentric outmoded fashion.

He worked tirelessly to collect the records of the past – indeed, all of these six classics existed in some form before his edition. Four works would later be added to the five Confucian classics that survived (the *Records of Music* was lost) to create a larger canon of Nine Confucian Classics.

Confucius's conservative philosophy championed the family unit as the basis for society, reinforced by respect for elders by their children, just as the elders venerated their ancestors and gave the same loving obedience to the Emperor that they expected from their own wives, and which his followers gave to the man they called 'The Great Sage' and 'The First Teacher'. His golden rule was 'Do not do to others what you do not want done to yourself.'

It is pleasing to note that, though the families of all the imperial dynasties of China have faded away, the Kongs (the descendants of Confucius) maintain the oldest, largest and most continuous genealogy in the world, currently mapping out eighty-three male generations since the death of the 'model teacher for ten thousand ages' in 479 BC.

6 ZOROASTRIAN IMMORTALS

Vohu-Mana ❊ Ash-Vahista ❊ Vairya ❊ Armaiti ❊ Haurvatat ❊ Amaratat

This list of immortals – the Amesha Spenta – is central to the ancient Indo-Aryan world-view propagated by the ancient Empires of Persia and maintained by the Zoroastrian community. Like so many six-strong lists, it can be crowned by a seventh – for they are in essence the servant/archangels of Ahura Mazda.

The Amesha Spenta are also seen as aspects of the supreme god of truth and light (Ahura Mazda) in his endless duel against Darkness and the Lie (Angra Mainyu) and his agents on earth. They have their areas of dominion, so that *Vohu-Mana* represents moral good purpose and is master of all animals; *Ash-Vahista* is righteous truth and lord of fire; *Vairya* is holy dominion and ruler of minerals and metals; *Armaiti* is holy devotion and lord of the earth; *Haurvatat* is divine unity and lord of the seas and water; and *Amaratat* is divine immortality and master of all plants and vegetables.

6 DAYS OF GENESIS

Light ✣ Firmament ✣ Land and Vegetation ✣
Heavenly Bodies ✣ Fishes and Birds ✣ Animals and Man

The story of the creation of the world was recorded in the book of Genesis in the seventh century BC by Jewish exiles working up their own ancient traditions under the strong influence of the 2,000-year-old urban culture of Mesopotamia. So, inevitably, it bears the strong imprint of Babylonian literary forms, albeit one imbued with an energy and conviction all its own – so much so that two similar but not identical tales are told in Chapter One and Chapter Two.

One of the more curious literary aspects of the story is the way of describing God like a king on his throne, command-ing his courtiers to 'Let there be' – and it happens, though due to the strict monotheism of the Jews there is no lower pantheon of gods to assist. This tale is familiar from older Babylonian texts which describe the creation of the world, notably the Enuma Elish and Atra-Hasis. The central device is the concept of the speaking of a thing (*logos*) being the necessary prelude to its creation.

The first three acts of Genesis are also depicted in the nature of a separation of the yet unformed but existing cosmos, such as the separating of darkness from light on day one, the waters above from the waters below in day two, and on day three the sea from the land. The next three 'creations' fill this universe with the different forms of life, concluding with the creation of man and woman, where the Genesis myth departs from its Babylonian models which have men and women as divine and equal (rather than woman as an adjunct of man – a concept in bizarre opposition to tens of thousands of years of veneration of the Mother Goddess).

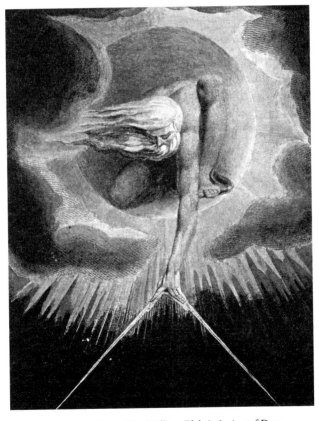

God creates the world in William Blake's Ancient of Days.

6 PATRICIAN FAMILIES OF ROME

Manlii (*gens Manlia*) ❋ Fabii (*gens Fabia*) ❋
Aemilii (*gens Aemilia*) ❋ Claudii (*gens Claudia*) ❋
Valerii-Cornelli (*gens Cornelia*)

The six major Patrician families of Rome – the *gentes maiores* – claimed descent from the priesthoods held by their ancestors at the time of the city's foundation by Romulus and the first seven kings, when the senate was just a gathering of priests checking that the royal decrees were consistent with the will of the gods. The *Manlii* remembered their origins from Etruscan Tusculum. *Fabians* claimed descent

Caligula – possibly not the finest hour for the Claudian patrician family.

from Hercules through Sabine highlanders and kept control of the ancient Lupercalia festival – though their detractors argued that their name derived either from 'peasant', 'bean' or 'ditch [cleaner]'. The *Aemilians* traced their origin to Sabine highland chieftains invited to Rome by the second king, Numa Pompilius, and their bloodline to Aemylos son of Ascanius – though others argued that they were descended from Romulus and Remus's sinful uncle, Amulius.

The *Claudians* were yet another Sabine family, 'distinguished by a spirit of haughty defiance, disdain for the laws and an iron hardness of heart', who were divided into either the very good or the very bad – and contributed the Claudian line of emperors (Tiberius, Caligula, Claudius and Nero) along with twenty-eight consuls, five dictators and seven censors. The *Valerians* had their own throne on the Circus Maximus and tended to ally with the Fabians to form a power block second in influence to the Cornelli.

The *Cornelli* were the most powerful of all the families, and it was said that one in every three of all the consuls of the Republic owed them some allegiance in blood. Their subsidiary clans included such powerful factions as the Scipio, Sulla, Lentulus, Dolabellae and Cinna families.

6 EVOLUTIONARY STAGES OF HISTORY

Clan communism ❈ Autocratic monarchy ❈ Feudalism ❈ Capitalism ❈ Socialism ❈ Communism

This is the Communist view of history, as set out by Marx and Engels, looking out over the wreck of the various social revolutions that were destroyed in the 1840s and dreaming of inevitable victory in the future. First, we have the primitive *clan communism* of hunter-gatherer families; then, once irrigated riverine agriculture is developed, the ancient

Engels and Marx charted a six-part theory of historical evolution.

autocratic monarchies, which endure as empires until they collapse from the weight of their own military-bureaucracy into the more enduring *feudalism*. With the growth of cities and maritime trading nations, feudalism matures into *capitalism*, which through the dictats of growth, decency and efficiency evolves into industrialised *socialism*, which perfects as *communism*.

6 MISHNAH OF THE TALMUD

Seeds ❋ Appointed times (festivals) ❋ Women ❋
Damages ❋ Holy things ❋ Purity

The Talmud, the great collection of Jewish oral tradition and textual commentary, was collated in the centuries after the fall of the Temple of Jerusalem and before the birth of Islam. It was created by more than 120 known *Tannaim*

(Rabbinical scholars) working over five generations. Their sixty-three chapters are arranged into the six orders (Shisha Sedartum) of the Mishnah – *Zeraim, Moed, Nashim, Nezikin, Kodashim* and *Tehorot* – as translated above.

6 PHYSICIANS OF ANTIQUITY

Plato ✻ Hippocrates ✻ Socrates ✻ Aristotle ✻
Pythagoras ✻ Galen

These six physicians were heroes of the medieval era, both to the Christian West and the Muslim East. Dante places them amongst the classical poets in the outer circle of Hell, which was set aside for virtuous pagans – a place of green fields overlooked by a castle with seven gates for the seven virtues.

6 ACTS OF MERCY

'For I was an hungry, and ye gave me meat: I was thirsty, and ye gave me drink: I was a stranger, and ye took me in: I was naked, and ye clothed me: I was sick, and ye visited me: I was in prison, and ye came unto me.'

As related in the Gospel of Saint Matthew (25:34–46) and still the only way to spot a true Christian.

6 LINES OF THE LAST DELPHIC ORACLE

The fair-wrought house has fallen ✻ No shelter has Apollo
✻ No sacred laurel leaves ✻ The talking spring is silent ✻
The voice is stilled ✻ It is finished

'It is finished' was traditionally believed to have been the very last utterance of the Delphic Oracle, sent in response to a question from the Roman Emperor Julian the Apostate. Julian was the descendant of Constantine the Great, who attempted to reverse the compulsory Christianisation of the Roman Empire. It is intriguing to note that the Oracle's last words were also spoken by Christ on the Cross.

THE 6 IMPOSSIBLE THINGS OF GLEIPNIR

Sound of a cat's footstep ❊ Roots of a mountain ❊
Beard of a woman ❊ Sinews of a bear ❊
Breath of of a fish ❊ Spittle of a bird

These rare ingredients were used by the dwarves commissioned by the Norse Gods to fashion a magical rope with which to bind Fenris the Wolf, one of the three monstrous sons of Loki. The rope, Gleipnir, was as thin as silk but stronger than any chain. Even Fenris couldn't break it, but instead bit off his hand to free himself.

5

THE QUINCUNX OF HEAVEN

Five has an elemental unity. It is the proof of Heaven, whether you count the petals of a flower or the lessons of divine love ticked off from the ends of your fingertips. There are five bodily senses, five colours, five directions, five organs, five limbs and five tastes. One of the most delightful pieces of sustained love for five-ness is Sir Thomas Browne's whimsical essay *The Garden of Cyrus* – a piece of English quietism constructed around the secrets of Heaven as revealed by a garden that ends with the warning: 'But the Quincunx of Heaven runs low ... and tis time to close the five portals of knowledge.'

THE PENTAGRAM – SOLOMON'S SEAL

The five-pointed star, the Pentagram, was a symbol of absolute authority to the Sumerian civilisation of Mesopotamia (modern-day Iraq), as early as the third millenium BC. It represented an additional axis of the royal authority reaching out to the four corners of the earth. Later, in classical Greece, it was used as a mystic symbol by Pythagoreans and in early Jewish lore it was associated with Solomon's Seal, a magical signet ring of King Solomon which gave him the power to command demons and speak to animals. (Confusingly, Solomon's Seal can also be depicted as a hexagon.)

This Seal of Solomon was revered by Jews, Christians and Muslims. Indeed an interlinked ribbon version – known as the Seal of Solomon – is used on the Moroccan national flag. Medieval astrologers interpreted the pentagram as a symbol for the five wounds of Christ. However, the symbol dropped out of Christian use, having been co-opted by medieval necromancers and modern witchcraft.

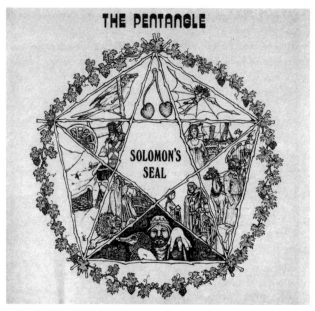

An English folk-rock classic from a hippie band keen on the mystical symbol. Goaty occult pentangles feature prominently in heavy metal.

Renaissance occultists made a distinction in the star's orientation. When pointed upwards the star was good, symbolising spirit presiding over the four elements of matter. Pointing down it was evil – the sign of the goat of black magic (whose face could be drawn in the star or its beard and horns suggested by the points). Wiccans have adopted the symbol (in its good form) as their emblem, and it is widely used by neo-Pagans, often as a pentacle, within an enclosing circle.

5 PRECEPTS TAUGHT BY BUDDHA

Self-effort ❋ Wisdom ❋ Alms-giving ❋
Meditation ❋ Pilgrimage

These five precepts accompany the road to Nirvana. In order to do the least harm on earth, one must also struggle to practise five abstentions: from the destruction of life, from theft, from falsehood, from sexual intercourse with anyone other than your acknowledged partner and from intoxicants which cloud the mind.

THE 5 PILLARS OF ISLAM

Profession of faith ❋ Alms-giving ❋ Daily prayers ❋ Fast
of Ramadan ❋ Pilgrimage to Mecca

As a young man travelling across the Islamic world and exhibiting an interest in their spiritual traditions, I was often given instances of how mankind was surrounded with the proofs of Islam, how the five fingers and the five senses could be used as a handy reminder of the five pillars of Islam, the five daily prayers and also remind one of the five prohibitions (pork, wine, gambling, adultery and divination). But the most charming evocation of five I ever came across was a scruffy old Moroccan shepherd, who plucked at flowers and even cracked open a cucumber to show how the world was ordered by five, which he explained was upheld by a verse of the Koran. I nodded politely at the time but years later came across Arberry's translation of the Sura al-Anam: 'Look upon their fruits when they fructify and ripen? Surely in all this there are signs for a people who believe.'

THE KHAMSA (HAND OF FATIMA)

The Hand of Fatima (or *khamsa* – literally 'five') is an expressive symbol: an open hand seemingly composed of three fingers and two opposing thumbs. It is ubiquitous in North Africa and the Middle East, used as a good luck pendant, a badge, an item of jewellery, a door knocker or woven into a hanging, repeated as a symbol in weaving or incorporated into calligraphy. It is commonly thought of as a charm against the evil eye – the often unwitting power to afflict bad luck with an envious glance.

The *khamsa* is commonly called the Hand of Fatima (though this seems to be Western rather than Arabic usage), Hand of

Five hands of Fatmima on a Moroccan amulet.

Mary, Hand of Miriam or, more anciently, the Mano Pantea of the Egyptian goddess Hathor, or the Hand of Venus. Its origins have been variously and contentiously traced back to phallic pendants (a cock flanked by a pair of testicles and a swirl of public hair), the open hand of the Buddha, the empty throne of Phoenicia, the Carthaginian goddess Tanit, the hanging lotus flower and Ishtar of Babylon. Perhaps its great virtue is that it remains a blank space, into which you can insert your own faith and symbolism. The symbolism of the open hand can be read by all cultures, as both blessing, salutation and a sign of peace – one which extends beyond humankind, for it is also acknowledged by animals.

5 BOOKS OF THE TORAH

Genesis ❋ Exodus ❋ Leviticus ❋
Numbers ❋ Deuteronomy

The five books of the Torah are collectively known as the Pentateuch, from their Greek translation made in third-century BC Alexandria. The Hebrew names for the holy books are derived from the opening sentences, so Genesis is *Bereshit* (In the beginning), Exodus is *Shemot* (Names), Leviticus is *Vayikra* (He called), Numbers is *Bamidar* (In the desert) and Deuteronomy is *Devarim* (Words).

All these five books are believed to have been written by Moses from the inspiration he received on Mount Sinai, though modern textual scholarship argues that a comprehensive rewriting occurred during the Jewish captivity in Babylon from the sixth century BC.

The Samaritan community believe that divine revelation stopped with these five books and that all other books in the Jewish canon are written by the hand of man, if not mere commentaries. They trace their historical descent from the northern kingdom of Israel that split off from the Kingdom of Judah after the death of King Solomon and have long worshipped God from an altar raised on Mount Gerizim.

5 COLOURS OF LUNGTA

Blue for space ❋ White for water ❋ Red for fire ❋
Green for wind ❋ Yellow for earth.

These are the colours seen in the wind-ripped Buddhist silk prayer flags that fly in Tibet and the mountain valleys of the Himalayas.

5 COMPONENTS OF THE SOUL IN ANCIENT EGYPT

Ren ❋ Ka ❋ Ib ❋ Ba ❋ Sheut

The simplest concept is *Ren*, which is literally your name: it lives for as long as you are remembered, or can be read about on inscriptions, or included in prayers for the ancestors and their achievements. *Ka* is also easy enough to translate into modern idiom, for it is that vital essence that makes the difference between the living and the dead, between life and dead meat, between a warm body and cold clay.

Ib is literally the heart, formed from a single drop of clotted blood extracted from your mother's heart at the hour of your conception or birth. By heart, the Egyptians meant not just the organ for pumping blood around your body, but the seat of your soul, the good directing force in your life, searching after truth, peace and harmony.

Ba is that which makes each of us unique and different, that which makes us strive and achieve, the motivator but also the hungry elemental force that needs food and sex. In

Images of ba from Dendera Temple.

some form, your *ba* is destined to survive after death, often depicted or imagined as a human-headed bird, which with good fortune will go forth by day to enjoy the light, but might also end up existing only in the dark, like the bat or the ruin-haunting owl. *Sheut* is your shadow, and by extension the other you, as well as being used to describe a statue, a model or a painting of a human.

5 CONFUCIAN BLESSINGS

Longevity ❊ Wealth ❊ Health ❊
Civility ❊ A natural death

The Five Blessings can be symbolised by a peach – a very auspicious Chinese symbol, linked with wishes for long life (often expressed by the number 10,000, with its suggestion of infinity and immortality). An image of nine peaches and five bats (linked to a peach because they sound similar) is therefore coloured with all sorts of suggestions about all these blessings.

THE 5 FAVOURED MEATS OF THE DESERT

Camel ❋ Rabbit ❋ Gazelle ❋ Chicken ❋ Mutton

As well as listing the sayings of the Prophet, his example in dress, manners and deportment was analysed and copied by his followers. And lists of five were very popular, as they tied in with the five pillars of Islam and the five daily prayers. These meats were all known to be eaten by the Prophet during his lifetime, though his favourite was shoulder of mutton (an attempt on his life was once made by poisoning mutton). There is also a list of his five favourite fruits: dates, melons, grapes, watermelon, cucumber.

5 FREEDOMS OF PSYCHOANALYSIS

Unimportant ❋ Irrelevant ❋ Nonsensical ❋ Embarrassing ❋ Distressing

Patients undergoing Freudian psychoanalysis must be free to say whatever comes into their head, however unimportant, irrelevant, nonsensical, embarrassing or distressing it might seem to be, and yet be sure of receiving the same quiet level of intent listening from their analyst.

THE ORIGINAL 5-MAN CABAL

Clifford ❋ Anglesey ❋ Buckingham ❋ Ashley ❋ Lauderdale

The term 'cabal' probably derives from secret gathering of Jewish mystics exploring the Kabbalah. But in England it was popularly believed that the first cabal was a discordant group of ministers who served King Charles II in pushing through a fairly murky alliance with the king of France. For the word

'Cabal' can be formed by the initial letters of the Lords Clifford, Anglesey, Buckingham, Ashley and Lauderdale.

THE 5 MANIFESTATIONS OF THE BUDDHA

Kakusandha Buddha ❊ Konagamana Buddha ❊ Kassapa Buddha ❊ Gautama Buddha ❊ Maitreya Buddha

These are the five principal Bodhisattvas: three lost to a mythic past, one historical founder and one Mahdi-like Messiah to come, 'the loving one', the future Buddha.

THE 5 CLASSICAL ORDERS

Doric ❊ Ionic ❊ Corinthian ❊ Composite ❊ Tuscan

The five orders of classical architecture form a familiar iconography of power in stone. Each of them look back to their origins in the simple wooden tree trunks, rudely shaped pillars of wood and bundles of reed and saplings of our Arcadian past.

Doric, Ionic, Corinthian, Tuscan and Composite.

5 POISONS OF OLD CHINA

Centipede ✵ Scorpion ✵ Snake ✵ Gecko ✵ Toad

The five poisons of old China could be combined into a charm to ward off evil and sickness. They also represent social evils – Confusion, Pride, Envy, Hatred and Desire – and on the unpropitious fifth day of the fifth month the hero-god Chung K'uei was displayed to keep them at bay. The poisons can also be linked to China's five elements – Wood, Fire, Earth, Metal and Water – and to the five directions of old China – East, South, Centre, West and North.

5 PRECEPTS OF APOLLO

As a child, learn good manners ✵ As a young man, learn to control your passions ✵ In middle age be just ✵ In old age, give good advice ✵ Then die, without regret

These precepts were received from the god Apollo at the oracle of Delphi and carved onto the funerary monument of Kineas (one of the founders of the Afghan city of Al-Khanum) by Clearchos, a disciple of Aristotle.

5 QUALIFICATIONS OF ISLAMIC VIRTUE

No honour is like knowledge ✵ No belief is like modesty and patience ✵ No attainment is like humility ✵ No power is like forbearance ✵ No support is more reliable than consultation

These qualifications were taught by Ali, the first male disciple of the Prophet Muhammad, his most valiant warrior, his cousin, son-in-law and father of his only male grandchildren. Ali was overlooked in the political succession to the leadership of Islam by the first three Caliphs, which allowed him to meditate on the essence of faith in Medina. He is the fountainhead of all the Sufi brotherhoods and mystical practices of Islam, and possessed all these virtues in abundance.

5 SUFI POWERS

Clairvoyance ❈ Remembering past incarnations ❈
Levitation ❈ Transmutation of elements ❈ Omniscience

When a mystical master experiences enlightenment, Sufi belief systems claim that he will also attain spiritual powers such as clairvoyance, remembering past incarnations, levitation, transmutation of the elements and omniscience. There are numerous popular stories about how Sufi, Buddhist and Zen masters can be at different places at the same time, and convocations of Hindu sages and Christian saints can magically assemble themselves together.

Sufi levitation in India.

5 RIVERS OF HADES

Acheron ❖ Cocytes ❖ Phlegethon ❖ Lethe ❖ Styx

Which is to say: the river of sorrow, the river of damnation, the river of fire, the river of oblivion and the river of hate, upon whose waters even the gods swore.

Some classical writers imagined Lethe as a pool of oblivion and added the pool of Mnemosyne (memory) beside it. Others envisaged flat, featureless misty land beside the rivers which they named the Fields of Asphodel. The Plain of Tartarus was reserved for more active punishment just as the Fields of Elysium or the Isles of the Blessed were reserved for blameless heroes. But even for such a proud hero-warrior as Achilles, it would be better to be the meanest ploughboy on its green earth than Emperor of all the Dead. That monarch was Hades Plouton – rich in lost souls and mineral wealth and married for all eternity to Persephone, the iron queen.

5 REGRETS OF THE DYING

I wish I had lived the life true to myself, not the life that was expected of me ❖ I wish I had not worked so hard ❖ I wish I could have expressed my feelings ❖ I wish that I had stayed true to my friends ❖ I wish that I had allowed myself to be happier

These are five modern regrets, as observed by an Australian nurse, Bronnie Ware, who specialised in caring for the dying in the last months of their life. She found that at the end of their lives people were possessed of a phenomenal clarity of vision, of which five themes surfaced again and again. Men were particularly prone to feeling that they had put all their passion into work and failed to attend to their children or their partners. *Carpe diem.*

5 RESTRAINTS ON SEXUAL PASSION

Do not violate the laws ❀ Do not disturb well-established customs ❀ Do not harm any of your neighbours ❀ Do not injure your own body ❀ Do not waste your possessions

These restraints were taught in the third century BC by the Greek philosopher Epicurus, despite his belief that 'a man never gets any good from sexual passion, and he is fortunate if he does not receive harm'. Epicurus was wise enough not to issue prohibitions and instead permitted his followers to 'follow your inclinations'. while establishing these provisions to discourage them.

5 SACRED MOUNTAINS OF CHINA

Tai Shan ❀ Song Shan ❀ Hua Shan ❀
Heng Shan (north) ❀ Heng Shan (south)

Five stands beside nine as one of the great organisational numbers within Chinese society. The five planets (excluding the sun and moon) were linked to the pentagram, to the five directions, the five elements, the five seasons, the five types of grain, the five tastes (Sour, Bitter, Sweet, Spicy, Salty) the five degrees of nobility, the five virtues (Benevolence, Knowledge, Faith, Righteousness, Propriety), the five street sounds (Calling, Laughing, Singing, Lamenting, Moaning), five smells (Goatish, Burning, Fragrant, Rank, Rotten), the five failings, the five disappointments, the five weapons, the five moral qualities, five weapons, five animals and five punishments. They could be arranged in fascinating charts and diagrams to make correspondences with each other, and the Five Sacred Mountains could be represented by the five Ancient Ones – symbols of every aspect of China and its material reality and spiritual associations.

5 STARS IN THE FLAG OF THE PEOPLE'S REPUBLIC OF CHINA

The large star stands for the Communist Party. The four smaller ones represent: the workers, farmers, soldiers and teachers; or worker, farmer, soldier and revolutionary student; or the Communist Party leading the way for the four compass points of the Middle Kingdom.

5 WIZARDS IN *LORD OF THE RINGS*

Saruman the White ❋ Gandalf the Grey ❋ Radagast the Brown ❋ Alatar also named Morinehtar ❋ Pallando also named Rómestámo

The Five are known as Wizards by men, and as the Istari by Elves., and their role is to assist Middle-Earth. Saruman is the man of skills; Gandalf is the elf of the staff; the dreamer; Radagast is the friend of birds and tender of beasts; Alatar (also named Morinehtar) and Palland (Rómestámo) are the sky-blue wizards who journey into the east and out of the story.

5 ORDERS OF TREE NYMPHS

Dryads ❋ Meliads ❋ Epimeliads ❋ Caryatids ❋ Heliads

The *Dryads* could be worshipped in oak forests; *Meliads* were associated with ancient ash trees; *Epimeliads,* with apple trees; *Caryatids* with the stately walnut groves; and *Heliads,* with all sorts of mossy ferns. Beneath them in power were the *Hamadryads*, who were so linked to the fate of their groves that they died when the trees were felled.

The tree nymphs were also mortal, but on a timescale that made them near immortals in the eyes of a man, as Hesiod has it, 'a chattering crow lives out nine generations of aged men, but a stag's life is four times a crow, and a raven's life makes three stags old, while the phoenix outlives nine ravens, but the rich-haired Nymphs outlive ten phoenixes.'

There were five other sorts of nymphs that a travelling shepherd might encounter – the *Nereids* and *Oceanids* of the sea; the *Naiads* and *Hydriads* of the river banks and lake waters. But it was the song of the *Epimeliads* that haunted him most, the divine watchers of the sheep.

5 REASONS FOR DRINKING

If all be true that I do think / There are five reasons we should drink / Good wine – a friend – on being dry / Or lest we should be by and by / Or any other reason why.

These lines are by Henry Aldrich (1647–1710), an archetypal Oxford man: a classical scholar, wit, philosopher, professor of logic, accomplished architect and Dean of Christ Church College. There are a number of versions, as its original was composed in Latin: *'Si bene quid memini / causae sunt quinque bibendi / Hospitis adventus / praesens sitis atque futura / Aut vini bonitas / aut quaelibet altera causa'*.

5 VIRTUES OF A TURKIC BEAUTY

An eyebrow like a willow leaf ❋ An eye like the kernel of an apricot ❋ A mouth like a cherry ❋ A face the shape of a melon seed ❋ A waist as thin as a poplar

Most of these are universal, though the Central Asian delight in a pair of eyebrows meeting like a slender bow on the brow of the beloved is a more particular taste.

5 WAYS THAT MUTTON CAN DISSAPOINT AN ENTHUSIAST

Ill-killed ❧ Ill-quartered ❧ Ill-cooked ❧
Ill-seasoned ❧ Ill-served

Dr Johnson's fivefold malediction suggests that London's eighteenth-century cooks left much to be desired.

Doubtless another night of disappointment for Dr Johnson (second right, with Boswell, left) in this painting of a dinner at Joshua Reynolds' house.

THE CANAANITE PENTAPOLIS

Sodom ❧ Gomorrah ❧ Segor ❧ Admah ❧ Seboim

This league of five ancient Canaanite cities in the Jordan valley has no exact location, which has encouraged persistent theories about them being buried below the waters of the Dead Sea. Apart from their intrusive style of entertaining passing travellers and the damnation that followed, it is also recorded in the Bible that four of the cities were utterly destroyed by the invasion of King Chodorlahamor. Segor alone survived, southeast of the Dead Sea; known as Zoar, it later sent a bishop to the Church Council of Chalcedon.

THE CINQUE PORTS

Sandwich ❋ Dover ❋ Hythe ❋ Romney ❋ Hastings

The confederation of the Cinque Ports – the five English ports closest to the continent of Europe – effectively managed and supplied a Royal Navy to English kings for 500 years, from their official incorporation in 1155 to their decline, when the Tudors could fund a Navy of their own. As the Cinque Ports grew in prestige and wealth, they added the two ancient towns of Rye and Winchelsea and eight confederate limbs: Lydd as a limb of *Romney*; Folkestone, Margate and Faversham as limbs of *Dover*; Deal, Brightlingsea and Ramsgate as limbs of *Sandwich*; and Tenterden as a limb of *Rye*. The addition of forty-two villages, hamlets and market towns to the league created a virtually self-governing dominion out of Kent. Aside from exemption from tax, tolls and the interference of royal justice, the Cinque Ports were duty-free and held rights over flotsam (wreckage), jetsam (goods knowingly cast on the waters) and lagan (underwater wrecks).

4

==

Fire ❊ Earth ❊ Air ❊ Water

This ancient division of the world of matter into a four categories underwrote a whole interlinked system of equiva-
lences that helped define human character, tend inbalances,
mend illness and peer into the future. For the four elements
were also assessed on a scale of hot and cold, wet and dry and
given particular associations.

Thus, *Fire* was both hot and dry and linked with one of
the four humours (the choleric) and the astrological signs
of Aries, Leo and Sagittarius. *Earth* was dry and cold, and
allied to black 'melas' bile (melancholic) and the three earth
signs of Taurus, Virgo and Capricorn. *Air* was both hot and
wet, and connected with blood and a sanguinous character
and the three air signs of Gemini, Libra and Aquarius. *Water*
was wet and cold, allied with a phlegmatic character and the
water signs of Cancer, Scorpio and Pisces.

The elements can also be allied to the four suits of cards,
either our modern symbols or the fourteenth-century forms

that are also used in the tarot pack: Cups (water), Swords (air), Batons (fire) and Coins (earth).

4 CARDINAL POINTS

North ❖ South ❖ East ❖ West

East and West have always been known to mankind as the places where the sun rises and where the sun sets. Indeed our very word 'east' is derived from the proto Indo-Aryan '*aus-to*', which means 'towards the sunrise'. Our obedience to the primacy of north (such as the arrow on the compass and the orientation of our maps) is a more recent shift. It is derived from '*ner*' ('down'). All the earliest maps, as drawn by the Chinese and Muslim cartographers, are orientated with south as 'up', just as the Emperor of China always sat on his throne facing south, towards harmony and prosperity.

The mystical writer John Michell examined how the sense of belonging to a point of the compass has brought out different natures in humanity. The north is the traditional land of warriors and iron-hard men, the east is the land of merchants and financiers, the south is the place for music, dance and emotional activity, and the west is the home of history, poetry and scholarship as well as the direction of enlightenment. It is curious how often this applies.

OVID'S 4 AGES OF CIVILISATION

Gold ❊ Silver ❊ Bronze ❊ Iron

According to Ovid's telling of history, our *Golden Age* was the time when Cronos ruled Heaven and when gods lived amongst mankind and noone laboured – that time when earth was populated by our hunter-gatherer ancestors. The *Silver Age* was when Zeus ruled over Heaven and mankind was taught agriculture and architecture – which we can tie to the inventions of the Neolithic, around 12,000 BC. The *Bronze Age* was the time of the first great wars, of temple- and empire-building but also of faith and order, which we can directly connect with the ancient civilisations in Iraq, Egypt, Anatolia, China and India. The *Iron Age* is our own time, when nation states were forged and mankind learned to mine, navigate, write and trade. But, as Ovid notes, mankind also became 'warlike, greedy and impious. Truth, modesty and loyalty are nowhere to be found.'

4 ARMS OF THE SWASTIKA

Sun ❊ Wind ❊ Water ❊ Soil

The four-armed swastika is one of the most universal symbols and can be found all over the globe, largely because it

Repeating swastika patterns from Tibetan decoration.

is a naturally occurring first feature in basket-weaving and mat-making. So swastikas have been unearthed in the earliest agricultural societies, be they in the Ukraine, China or on the banks of the Indus. The Greeks knew the symbol as a *tetraskelion* (four-legged) or *gammadion* (from the arm shape of the letter gamma) and its frequent use as part of a repetitive pattern makes the link to the recurring cycle of the four seasons very strong. It was associated with such benign goddesses as Demeter, Ceres, Hera and Artemis. The word 'swastika' is derived from Sanskrit and means either 'well-being', 'to be good' or refers to a 'being with a higher self'. In India it is associated with Vishnu and Shiva and revered by the Jains. In China it is a Taoist symbol for divine power and in Buddhism it is one of the auspicious signs on the Buddha's foot. So it is a bizarre irony that, since its adoption by the Nazi party in Germany in 1920, it is now associated with racism, ignorance, genocide and totalitarianism.

THE 4 SUITS OF A PACK OF CARDS

Clubs ✳ Diamonds ✳ Hearts ✳ Spades

If you count up the numerical value of a whole pack of cards – reckoning on 11 for a Jack, 12 for a Queen and 13 for a King – you reach 364, which with the addition of one for the joker makes 365, the number of days in the year. The four suits can also be read as symbols of society and human energy: *Clubs* representing both the peasantry and achievement through work; *Diamonds,* the merchant class and the excitement of wealth creation; *Hearts,* the clergy and the struggle to achieve inner joy; *Spades,* the warrior class institutionalised into the nobility and the fractious problems of life.

The pack of cards came to Europe sometime in the fourteenth century, imported by Italian merchants who discovered their use during trading missions to the cosmopolitan cities of Mameluke Egypt. The symbols they imported – swords, batons (or wands), cups, and coins (or rings) – are still used in Spain, Greece, Portugal and Italy. The modern four suits seem to have evolved in France, specifically Paris and Rouen, in the late fifteenth century and were quickly taken up by the English. The French also added the concept of the Queen, for initially the court cards were based on the sequence of King, Cavalier and Servant – or, as the original Mameluke Egyptians had it, Malik (King), Naib Malik (Viceroy) and Thaim Naib (Deputy). The triumph of the Ace was another French innovation, traditionally added after the Revolution in honour of the rabble toppling the King.

The Egyptians themselves seem to have developed the pack of cards from China, where numerically printed sheets grouped into four divisions can be traced back to the concubines of the Tang dynasty (618–907).

THE 4 CARDINAL VIRTUES

Wisdom ❉ Bravery ❉ Temperance ❉ Justice

Plato considered that the ideal state should be governed by 'the wise, brave, temperate and just'. These virtues – often listed as Prudence, Fortitude, Temperance and Justice – were popularised by Christian apologists and combined with the three theological virtues (Faith, Hope and Charity) to create a group of seven virtues to stand in opposition to the Seven Deadly Sins. *Wisdom* (or its feminine archetype, Prudence) is often depicted with a book, mirror, snake and compass. *Bravery* (fortitude) may be found standing next to a Greek helmet, a spear, shield, Samson's pillars or a Herculean club and Nemean lion skin. *Temperance* may be spotted holding a sheathed sword, a torch, a clock or mixing water into wine. *Justice* remains a familiar modern figure with her blindfold, an upheld sword in one hand and a pair of scales in the other.

THE 4 GAMES OF ANCIENT GREECE

Pythian ❉ Isthmian ❉ Nemean ❉ Olympic

The *Pythian Games* were held in honour of Apollo at Delphi; the *Isthmian Games,* at Corinth for Poseidon; the *Nemean Games,* at Nemea in honour of Zeus. But the most famous in the ancient world were the *Olympic Games*, held in honour of Zeus at Olympia in the Peloponnese, which attracted city states from across the Greek (and later Roman) world.

Each of these PanHellenic Games were held at intervals of either two or four years and were arranged so that each year there was at least one competition open to any free-born Greek. The first recorded winner is from 776 BC though the practice was considerably more ancient. The Games seem to have been designed to select the very best in order that they could offer up a sacrifice that would be the most pleasing to the Gods. This also explains the sacred truce, the cult of heroic nudity, the simple garlands awarded to the victors and the decision of the Christian Emperor Theodosius to close down the games in 394 BC.

4 PLEASURES OF OMAR KHAYYAM

Here with a Loaf of Bread beneath the Bough / A Flask of Wine, a Book of Verse – and Thou / Beside me singing in the Wilderness / And Wilderness is a Paradise now.

The medieval Persian poet and philosopher Omar Khayyam was introduced to the West by a Victorian poet called Edward Fitzgerald, who produced a very loose translation of his *Rubaiyat*, or 'quatrains'. His book had huge popularity in the mid-nineteenth century and even led to Khayyam's rediscovery in Iran. Fitzgerald took what he wanted from

Wilderness as paradies: Edmund Sullivan's illustration from Edward Fitzgerald's Victorian edition of the Rubaiyat of Omar Khayyam.

the original verse, lending it a romantic humanism, beneath which glow the truths of the Sufi.

4 CASTES

Brahmin ❖ Kshatriya ❖ Vaishya ❖ Shudra

In traditional Indian society, *Brahmin* are the priests, *Kshatriya* are the warriors, *Vaishya* are the merchants, and the *Shudras* are the labourers of traditional Hindu society. (A fifth caste is the so-called Untouchables or *Dalit*.) The four castes are comparable to the four estates of medieval European society: the Clergy, Nobility, Merchants and Peasants.

4 DEGREES OF ATTACHMENT

Secure ❖ Anxious and Preoccupied ❖
Avoidant and dismissive ❖ Disorganised

These human characteristics, which can already be assessed by the time a child is 18 months old, are based around four major observational themes: Proximity Maintenance, Safe Haven, Secure Base and Separation Distress. At their root they are but measures of the successful exchange of comfort, warmth and pleasure between an infant and its parents that was first conceived by Sigmund Freud and greatly extended by the work of John Bowlby.

4 TYPES OF CAVIAR

Beluga ❖ Sterlet ❖ Osetra ❖ Sevruga

Caviar is the edible squishy eggs (roe) of the sturgeon, a slow-moving, bottom-grazing fish that can grow to twelve feet in length. It was originally associated with the Caspian Sea but is now bred in other regions of the world due to the fantastic price that caviar fetches and the decline in sturgeon numbers in the polluted inland sea. *Beluga* is the most expensive variety, composed of large, soft, pea-sized eggs (normally packed into a blue tin); *Sterlet* is small and golden coloured (golden tin); *Osetra* is medium-sized, from grey to brown (yellow or green tin); while *Sevruga* (red tin) are the small black and grey eggs.

4 CONTINENTS

Europe ❋ Asia ❋ Africa ❋ America

The four continents might seem to fit into the ancient sacred categories of 4 very neatly, but the concept is comparatively recent. Before the maritime discoveries of the fifteenth and sixteenth centuries, the fundamental concept was of three continents – Asia, Europe and Africa – knitted together by such lynch-pins of the world as Jerusalem and Constantinople, Antioch or Alexandria.

4 MAGICAL POSSESSIONS OF ODIN

Draupnir ❋ Gungnir ❋ Sleipnir ❋ Mimir

The old Norse god Odin had a quartet of magical possessions. *Draupnir* was his gold ring, which every ninth night spawned another eight rings. *Gungnir* was a sacred spear that never missed its target. *Sleipnir* was an invincibly fast horse with eight legs. *Mimir* was the talking head with which he could look into the future, the knowledge of which first gave him joy, then sorrow, then grief.

4 EVANGELISTS

Matthew ❋ Mark ❋ Luke ❋ John

The Four Evangelists are often depicted, in fresco or mosaic, in the corner squinches that support the central dome of a church. *Saint Matthew*, or rather the Gospel he wrote, is symbolised by a winged angel; *Saint Mark,* by a winged lion; *Saint Luke,* by a Bull or a winged ox; and *Saint John* by an eagle.

Matthew is referred to as Levi in the Gospels of both Mark and Luke and also gets a mention in the Gospel of Thomas. He was the son of Alphaeus, a tax collector at Capernaeum who was summoned to become a disciple when Jesus bade him 'Follow me'. The name 'Matthew' is from the Hebrew for 'gift of love'. He established many Christian communities and was martyred in either Ethiopia, Parthia or Kyrgyzstan.

Mark was a cousin of Saint Barnabas, with whom he travelled on one of the first missionary journeys with Saint Paul. His Gospel is symbolised by a winged lion, recalling both Christ as the royal lion of the tribe of Judah and as a symbol for resurrection. He established the church in Egypt over which he presided as the first Patriarch of Alexandria. A tradition records that he retired to an empty valley tucked into the Cyrenaican hills (of eastern Libya) to write his Gospel.

The ox of *Luke* stresses the sacrificial nature of Christ. By tradition Luke was a physician and a follower of Saint Paul, who preached in Egypt and Greece. He painted the first icon, a portrait of the Virgin, as well as a vision of the Christ Child. He may also have been the editor who put together the Acts of the Apostles. He was crucified with Saint Andrew at the port of Patras in southern Greece.

Saint John the Beloved Disciple, John the Divine and John the Evangelist may all be one and the same man. If so, he was the one disciple to escape a violent death. John was also the one disciple who witnessed the Crucifixion, alongside the women of Christ. After the Crucifixion, John took the Virgin to Ephesus but he was later banned from the city for disturbing the public order with his miracles. In exile on the island of Patmos, he dictated his visionary Gospel in a cave.

4 HORSEMEN OF THE APOCALYPSE

White ❊ Red ❊ Black ❊ Pale

In the Book of Revelations of Saint John, after the lamb of God breaks open the first four of the seven seals, four beings are summoned forth that ride out on white, red, black and pale horses. They usher in the beginning of the end of the world. The *white horse*, with its rider holding a bow and a crown, has variously been taken to mean Caesar, the Par-

The Four Horsemen of the Apocalypse in a woodcut from the Cologne Bible of 1479.

thian Emperor of the East, or even Christ himself set upon the road of conquest. The *red horse* and its rider equipped with a large sword is usually taken to mean war. Famine with his measuring scales rides a *black horse*, while death is mounted on the sickly *pale horse*. Like much of the Book of Revelation, this vision can be linked to a passage within the Old Testament, in this case to the Prophet Zachariah's reference to four horsemen, though they are patrollers rather than agents of destruction.

The image of the four destructive horsemen has had enormous resonance in religious art over the centuries, most famously in Dürer's engravings. And they had a big revival in modern popular culture, featuring in fantasy literature (notably Terry Pratchett's *Discworld*), comics (as Marvel supervillains and Ninja Turtles), music (Coldplay's 'Viva la Vida' is one among scores of songs – including a large heavy metal contingent) and videogames (most recently, and successfully, the horsemen have a leading role in *Call of Duty: Modern Warfare*).

TETRAMORPHS

Angel ✳ Lion ✳ Oxen ✳ Eagle

These are the four figures that support the throne of Heaven, as revealed by the Prophet Ezekiel and further developed in Saint John's Book of Revelation. They seem closely allied to the four protecting female deities of Egypt – Isis and Nephthys, Hathor and Sekhmet – who watch over the infant god Horus in the Egyptian mythic tradition. The imagery is certainly beguilingly familiar, for there are two winged figures (the sister goddesses Isis and Nephthys) depicted with outstretched wings, while Hathor is the great cow-headed, nurturing Mother Goddess, and Sekhmet the fierce and brave lion-headed goddess.

4 GREATER PROPHETS

Isaiah ❋ Jeremiah ❋ Ezekiel ❋ Daniel

The four greater prophets of the Old Testament were favoured in the Middle Ages as precursors of Christ. They can often be seen amongst the attenuated figures holding scrolls on the great west fronts of a Gothic cathedral. *Isaiah* can be identified by his great beard and sometimes holds a saw with which he was martryed; *Jeremiah* holds Christ's cross; *Ezekiel*, a double wheel in allusion to his vision; and *Daniel* is normally associated with a lion or a fiery furnace.

The Islamic tradition also has a list of four prophets who can be appealed to in time of need. They number *Idris* (Henuh from the Old Testament); *Issa* (Jesus, who in Islamic tradition was not crucified but ascended to Heaven), *Elijah* (who was taken up to Heaven in a fiery chariot and is prophesied to return: 'I will send you Elijah the prophet before the coming of the great and terrible day of the Lord'); and most mysterious, *el-Khidr*, 'the green one', who is like a Muslim Saint George, going about his inscrutable tasks.

4 BUDDHIST PILGRIMAGES

Lumbini ❋ Bodh Gaya ❋ Sarnath ❋ Kusinara/Kusinagar

The Empress Helena, who built shrines to attract Christian pilgrims to the holy places, may have been consciously following the tradition of the Emperor Ashoka, who had done the same for the Lord Buddha in the third century BC. Some 250 years after the Buddha had left this world, the emperor, a convert to the religion, built a canopy supported by four pillars to mark the 'diamond throne' where he had sat under a giant fig tree and meditated his way to enlightenment.

The Buddhist pilgrimages are to *Lumbini*, in modern Nepal, the birthplace of Siddhartha Gautama; to *Bodh Gaya* in the modern Indian province of Bihar, the Buddha's place of enlightenment; and to *Sarnath* and *Kusinara/Kusinagar*, both in Uttar Pradesh, where he first taught and later expired.

4 SYMBOLS OF BUDDHIST ENLIGHTENMENT

Riderless horse ❊ Bo or Bodhi tree ❊
Empty throne ❊ Great wheel

These emblems were especially popular as rock carvings in the early period of Buddhism when human representations of the Buddha were not made, let alone worshipped.

4 WORDS OF BUDDHIST CHANT

Omm ❊ Mani ❊ Padme ❊ Hum

This is the universal Buddhist mantra: 'Hail to the jewel in the heart of the lotus.' The exact meaning, power, efficacy and use of the chant is a complex subject, for it has mutiple references that elide into every aspect of Buddhism, but it is a common tradition to break it down into its six Sanskrit syllables, which equate with a spectrum of six colours, six wordly deities, six deities, six realms of existence and six vices (which can be kept under some control by the chanting of the six syllables). So under this simplistic reading: *Om* fights against pride but opens the doors to generosity and is coloured white; *Ma* struggles against lust, nurtures compassion and is green; *Ni* resists desire, encourages patience and is yellow; *Pad* dispels ignorance, enhances diligence and is blue; *Me* struggles against possessiveness, enhances renunciation and is red; while *Hum* counters aggression and anger, strengthens wisdom and is black.

4 VOICES AND STRING QUARTETS

Soprano ❋ Alto ❋ Tenor ❋ Bass

The four voices required by a chorus are (descending in pitch) soprano, alto, tenor and bass. The voices have their instrumental counterparts in the string quartet – one of the abiding images of Western high culture, as if a group of four musicians can together aspire to express something beyond our humanity. It was Mozart (who wrote twenty-one string quartets) who perfected the form, using violin, two violas and a cello, though some argue that with the addition of a third viola and the composition of his six string quintets he created his most authentic voice.

THE QUADRIVIUM

Arithmetic ❋ Music ❋ Geometry ❋ Astronomy

The Quadrivium ('Four Paths') was the traditional division of medieval learning. It was first defined in the sixth century AD and dominated the intellectual imagination of Europe for the next thousand years. It was not a complete course in itself, just the middle passage of learning, began after the Trivium (Grammar, Logic and Rhetoric) had been achieved to reach the symbolic number of the mastery of seven liberal arts. Its subjects are all studies of number: the pure numbers of *arithmetic*, *music* as number in time, *geometry* as number in space, and *astronomy* as number in both space and time.

Once the Quadrivium had been mastered, a student was in a position to tackle the two most revered subjects of philosophy and theology, or to drift off into such practical fields of study as medicine, architecture or law.

4 HOLY MARSHALS OF GOD

Quirinus of Neuss ❊ Saint Hubert ❊
Saint Cornelius ❊ Saint Anthony the Great

The cult of the Four Holy Marshals of God developed in the medieval Rhineland, collecting together a college of saints who could be appealed to for protection against diseases and infections, and for the welfare of animals. *Quirinus* was efficacious against smallpox and useful with horse ailments; *Hubert* was the master of hounds and also useful against rabies and wild dogs; *Cornelius* looked after cattle, cramps and epilepsy; and *Anthony the Great* (a Coptic monastic saint from Egypt) could care for pigs as well as keep the plague at bay.

Anthony the Great – handy for pig husbandry.

4 CELTIC SEASONAL FEASTS

Winter ❧ Spring ❧ Summer ❧ Autumn

The four seasons were celebrated by the Celts of ancient Britain in the festivals of *Samhaim* (1 November), *Imbolc* (1 February), *Beltene* (1 May), *Lughnasa* (1 August) – or in their modern dress, Guy Fawkes Day, Saint Valentines Day, May Day and the August Bank Holiday.

4 DOCTORS OF THE CHURCH

Ambrose ❧ Jerome ❧ Augustine ❧ Gregory

After Saint Paul, no four men had so great an influence on the early Christian church as the 'Doctors', who are depicted in thousands of frescos, sculptural reliefs and wall paintings, though it must be remembered that they affected only the western, Latin half of Christianity, not the old Eastern homeland of the faith. There, the Orthodox communion venerate instead the Four Greek Fathers: Saints John Chrysostom, Basil the Great, Athanasius and Gregory Nazianzen, who can often be spotted on the lower register of church walls in the apse.

In the Latin panoply, *Saint Ambrose* (c.339–397) was a bishop of Milan. He is often associated with the whip with which he corrected heretics and a beehive to represent his honeyed preaching. *Saint Jerome* (c.342–420) was the creator of the Latin Bible; he is depicted as a scholar hermit with a cardinal's hat in his hut alongside the lion who befriended him after he had plucked out a thorn from his paw. *Saint Augustine* (354–430), author of *City of God*, founded the ideal of the Church as the heir of the fallen power of Rome. (In his youth, he had uttered the famous prayer, 'Lord, Grant me chastity and continence, but not yet.') *Pope Gregory I, the Great* (c.540–604), whose multitude of missions included evangelising the pagan Anglo-Saxons, is usually depicted with his papal tiara and pontifical cross.

4 HUMOURS

Sanguine ❖ Choleric ❖ Melancholic ❖ Phlegmatic

The Four Humours or Temperaments were a foundation of European medieval medical philosophy. The ideal was for a balance of the four, which were conceived to be based on the properties of blood (*Sanguis*), yellow bile (*Kholê*), black bile (*Melas*) and Phlegm in the body.

A predominance of *Sanguine* was believed to create an easy-going, sociable, pleasure-seeking type of person. A *choleric* character was fiery, strident and ambitious. *Melancholic* was watery and emotional and created thoughtful, introverted and intellectual types. *Phlegmatic* was slow and earthy but also governed the most relaxed, content and quiet of types.

4 LEGIONS THAT CONQUERED BRITAIN

II Augusta ❖ IX Hispana ❖ XIV Gemina ❖ XX Valeria

The *II Augusta* legion was raised by Octavian to fight in the civil wars and won battle honours in Spain and on the German frontier before joining the army that conquered Britain in 43 AD. Despite a troubled time during Boadicea's 60 AD rebellion, it remained one of the principal garrison legions of Britannia. The *IX Hispana* was raised by Pompey, fought under both Caesars but was destroyed in 161 fighting the Parthians in the east. The *XIV Gemina* was raised by Caesar for the conquest of Gaul, fought all over Britain in the years after the conquest and won the title 'Martia Victrix' during

Boadicea's revolt before being moved back across the Channel to garrison the Batavian frontier. The *XX Valeria* was the British legion par excellence, with its garrison headquarters at Chester and the boar as its emblem. Raised by Augustus, it won its name and title 'Valeria Victrix' during the conquest of Yugoslavia, after which it was selected by the Emperor Claudius for his conquest of Britain.

4 RIGHTLY GUIDED CALIPHS

Abu Bakr ✳ Omar ✳ Uthman ✳ Ali

These names, written in wonderfully flowing script, are often placed in medallions to ornament a Sunni mosque – most famously on the four squinches that uphold the dome of Ayia Sophia in Istanbul.

Abu Bakr, the father of the Prophet's beloved wife Aisha, ruled as Caliph from 632 to 634 AD. *Omar* (or Umar) was the father of Hafsah (the Prophet's most intellectual wife) and ruled from 634 to 644. *Othman* (Uthman), twice son-in-law of the Prophet, ruled as the third Caliph from 644 to 656. *Ali*, adopted son, paternal cousin, first disciple and son-in-law of the Prophet, was the fourth Caliph in the Sunni tradition (656–661) but also the first Imam of the Shiites (632–661).

4 RIVERS OF MOUNT KAILASH

Ganges ✳ Indus ✳ Sutlej ✳ Brahmaputra

Mount Kailash is holy to a fifth of mankind – Hindus, Jains and Buddhists – as well as being the source of the sacred Mount Meru at the dawn of Aryan consciousness. It is the source of four holy rivers. The *Ganges*, the sacred river of

India, flows through the gorge of the Peacock's Mouth. The *Indus* is the Lion-Mouthed river which gave birth to the Harappa, one of the oldest, most peaceful and little-known of the world's urban civilisations. The *Sutlej* flows through Elephant's Mouth gorge and the *Brahmaputra* through Horse Mouth gorge.

THE 4 RIVERS OF PARADISE

Pishon ✢ Gilion ✢ Hiddikel ✢ Euphrates,

From the roots at the foot of the Tree of Life flows a spring, the source of the four rivers of paradise which flow to the cardinal points to form a cross. Listed as Pishon, Gilion, Hiddikel and Euphrates, they are often used as a Christian metaphor for the Gospels or the redeeming Crucifixion. *Pishon/Fison* is often placed in Southern Arabia; *Gilion/Gellon* usually interpreted as the Nile; and *Hiddekel* as the Tigris, the river which Dionysus, on his sacred journey through the East, crossed by riding on the back of a tiger, having built a bridge across the *Euphrates* (which runs through Syria and Iraq).

4 SONS OF HORUS

Imsety ✢ Qebsennuef ✢ Duamutef ✢ Hapi

The heads of the four sons of Horus can be found guarding all the tombs of ancient Egypt, for they decorated the tops of the Four Canopic jars that held the internal organs of the dead. *Imsety*, the human-headed guardian of the South, has charge of the liver; *Qebsennuef*, hawk-headed guardian of the West, has charge of the intestines; *Duamutef*, jackal-headed guardian of the East, has charge of the stomach; and *Hapi*, ape-headed guardian of the North, has the lungs.

4 WOODS OF THE CROSS

Cedar ❁ Cypress ❁ Palm ❁ Olive

Tradition has it that the upright post of the Cross was fashioned from *cypress* inserted into a *cedar* post embedded in the earth, the crossbeam was made from *palm* and the tablet upon which was written (in Hebrew, Greek and Latin) 'Here Dies the King of the Jews' was made from *olive*.

4 SUFI QUESTIONS

How did you spend your time on earth? ❁ How did you earn your living? ❁ How did you spend your youth? ❁ What did you do with the knowledge I gave you?

This is a traditional Sufi teaching about the passage of the soul after death, which is ushered before the throne of God and asked just these four questions. I first saw it on a poster in the office of a Moroccan travel agent in Tangiers, but having failed to remember it properly was delighted to stumble across it thirty years later in Elif Shafak's novel *Honour*.

3

DIVINE TRINITIES

Religions are obsessed by trinities. The Greeks had the Olympian trinity of Zeus, Athena and Apollo; the Romans, the Capitoline triad of Jupiter, Juno and Minerva; while Egypt's great Nile trinity was Osiris, Isis and Horus. Further back in time, in Sumeria, the Great Goddess could be addressed in the forms of Asherah, Astarte and Anat, just as in the West she could be Diana-Nemorensis, Phoebe-Selene and Hecate-Persephone. Selene herself is bound up with the triple identity of the three groups of Fates (Moirai), Furies (Erinyes) and Graces (Charites). And Hecate was often depicted as a triple-headed woman.

Little surprise, then, to find Christians shaping their supreme deity around the Trinity of Father, Son and Holy Spirit. Or the biblical Jesus's three-part symbolism: rising from the dead on the third day, being denied thrice by Saint Peter, and crucified amid three crosses. Nor that Muslims regard holiness on this earth as having been possessed for just three generations (the Prophet, followed by his cousin and son-in-law Ali and then his two grandsons, Hassan and Hussein).

TRICOLONS

Tricolons are a rhetorical flourish – a sonorous list of three concepts, often escalating in significance. The most famous is Julius Caesar's proud despatch to the Senate of Rome following his expedition to the near-mythical, mist-clouded Isle of Britain: *'Veni, Vidi, Vinci'* ('I came, I saw, I conquered'). But Caesar's tricolon is run close by those great orators Lincoln and Churchill, while in recent years Barack Obama has revived the form, sometimes going for a double tricolon, as in this speech echoing the Declaration of Independence:

'Our generation's task is to make these words, these rights, these values – of life, liberty and the pursuit of happiness – real.'

Here are some of the all-time classics:

'Government of the people, by the people, for the people.'

The threefold manifestation of a fully functioning democracy as defined by Lincoln. He also, apparently in casual conversation, made a masterly analysis of the limits of the dark arts of political life:

'You can fool some of the people all of the time, and all of the people some of the time, but you cannot fool all of the people all of the time.'

Churchill was an enthusiast for the tricolon, promising

'Blood, sweat and tears'

as all that he could offer the people of Britain if they were to follow him in offering uncompromising opposition to Nazi Germany. It was matched only by his tricolon of praise for that handful of gallant knights of the air who defended the shores of Britain:

'Never in the field of human conflict was so much owed by so many to so few.'

But perhaps most glorious of all is the inscription on the Statue of Liberty at the entrance to New York, taken from a sonnet by Emma Lazarus:

'Give me your tired, your poor, your huddled masses yearning to breathe free.'

3 FATES

Clotho ❋ Lachesis ❋ Atropos

In the classical world, it was the three white-robed Fates who spun, measured out and cut the thread of life: *Clotho* spins, *Lachesis* measures and *Atropos* cuts. They were known as the Moirai to the Greeks – those who 'apportion' your time – and by the superstitious Romans by the euphemism of Parcae, 'the sparing ones'.

3 FURIES

Megaera ❋ Tisiphone ❋ Alecto

The Furies (*Erinyes*) were also known in Athens by the cautious euphemism of the *Eumenides* – 'the kindly ones' – and were worshipped in a cave below the Parthenon. *Megaera* the jealous, *Tisiphone* the blood avenger and *Alecto* the unceasing were elemental figures of female power – the relentless winged spirits of Conscience, Punishment and Retribution hunting down the guilty.

Later traditions identified them as the daughters of Gaia, inseminated when the bloody testicles of the ancient god Uranus, who was castrated by his son Cronus, fell to the earth. They are sometimes depicted like a winged spirit of victory, sometimes like a Medusa figure dripping with gore and a scalp sprouting a mane of thrashing serpents.

3 GRACES

Aglaea ❊ Euphrosyne ❊ Thalia

In ascending order of age we have *Aglaea* (Splendour), *Euphrosyne* (Mirth) and *Thalia* (Good health or happiness). The three sisters have been obsessively painted and sculpted for thousands of years as the embodiments of beauty, charm and creativity. As the benign face of the ancient triple goddess, even Hesiod and Homer can seem vague about their origins, and so there are conflicting stories of them being the daughters of either Aphrodite, Apollo, Zeus or Dionysus. Like the Furies, their chapel stood in the caves around the Acropolis, where ancient mysteries were performed.

The Three Graces from a fresco uncovered at Pompeii.

3 GORGONS

Stheno ❈ Euryale ❈ Medusa

Stheno (the mighty), *Euryale* (the far-springer) and *Medusa* (the queen) were, again, ancient aspects of the triple goddess in her destructive, vengeful form, though they were later demoted in scale to malevolent creatures. Perseus's murder of Medusa can be read as a mythic explanation of the toppling of the old female-ruled universe by a new breed of priest-warriors. However, the power of the old beliefs doesn't wane easily: Medusa's blood turned into serpents when it penetrated the ground, and gave birth to the winged horse Pegasus when it met the sea.

3 GODDESSES OF PRE-ISLAMIC ARABIA

Al-Lat ❈ Al-Uzza ❈ Manat

These goddesses were worshipped at a trio of shrines around Mecca until they were destroyed by the triumph of Islam, an event directly attested to in the Koran (Surah 53, 19–22). Nabataean inscriptions also make reference to this trinity, naming *Al-Lat* as the Mother Goddess of prosperity, *Al-Uzza* as the mighty one and *Manat* as 'the other', 'fickle' or 'fate'.

THE TRIPARTITE REVOLUTIONARY VIRTUES

Liberty ❈ Equality ❈ Fraternity

Nothing has ever quite matched the elan of idealism expressed in these tripartite watchwords of the French Revolution, which became the national motto of the nation. They are attributed to a Parisian printer, Antoine-François Momoro, though at the time of the Revolution there were several variants, and lists might include *Amitié* (Friendship), *Charité* (Charity) or Union – and there was often a qualifier – *'ou la Mort'* (or death). The latter was discreetly dropped after the Reign of Terror.

3-LEAVED SHAMROCK

The Celts of Western Europe had *Matronae* (Mother Goddess) shrines that could multiply from three to six to nine identities, a belief shared in pre-Christian Ireland, where the eating of the sorrel was a form of divine communion with the goddess in the spring. Saint Patrick cleverly co-opted the shamrock as proof of his new biblical version of the trinity.

CLOGS TO CLOGS IN 3 GENERATIONS

One of the fundamental historical ideas behind the works of that genius medieval historian Ibn Khaldoun is the natural decline of dynastic energy. To shorten his great work to a sentence: the Questing-Conqueror unites his people behind some cause, is succeeded by the Builder-Administrator, and then the house of power disintegrates amongst a feuding family of spoiled young princelings. In Lancashire this rise to wealth and its descent again to poverty is charted in the Victorian phrase 'from clogs to clogs in three generations'.

3 GOOD THINGS OF THE CALIPH'S DREAM

Intellectual achievement ❋ Justice ❋ Public welfare

In this medieval tale, Islam absorbs the teachings of Socrates. The all-powerful caliph has a dream encounter with a philosopher of very unprepossessing looks, 'ruddy complexion with joined-up hairy eyebrows, a bald head and bloodshot eyes reclining on a divan.' The caliph asks him to define three things that are good. He responds: 'Intellectual achievement; justice; public welfare.' To which the caliph replies: 'But what else! What else?' At which he receives the reply, 'There is nothing else.'

3 VIRTUES

Faith ❋ Hope ❋ Charity

The Three Virtues are defined by Saint Paul in the Book of Corinthians: 'And now abideth faith, hope, and love, even these three: but the chiefest of these is love.' In Christian imagery, *Faith* is depicted with a cross and chalice, *Hope* is shown with her hands in prayer and reaching up towards a Heavenly crown (sometimes with maritime associations like an anchor) and *Charity* is seen as open-handed. They are sometimes linked with the threefold symbolism of the growth of a soul: External, Internal, Supreme.

The Caliph Uthman (a companion of the Prophet) defined the three virtues of a Muslim more specifically as:

To feed the hungry ❋ To clothe the naked ❋
To read and teach the Koran

3 PRINCIPLES OF A GOOD ZOROASTRIAN

Humata ❋ Hukhta ❋ Havarshta

Humata (good thoughts), Hukhta (good works) and Havarshta (good words) will all assist true believers after death when they cross the bridge of *Chinvat* (judgement) to either a Paradise of Songs or fall to the House of the Lie.

FREUD'S 3 ELEMENTS OF PERSONALITY

Id ❋ Ego ❋ Superego.

Sigmund Freud conceived of the personality as consisting of three interrelated influences. The *Id* is a person's natural instincts and desires, such as to procreate, to eat and to survive. The *Ego* uses reason to mediate between reality and the Id, so one might say that in today's world I can only afford two children, or there are six people needing to eat so I can't have the whole chicken. Lastly, there is the *Superego*, akin to the conscience, and thought to originate as an internal version of what parents, school and society teach. This introduces the concept of 'I should' – for example, share my good fortune with those less fortunate than myself.

3 KABBALISTIC COMPONENTS OF A SOUL

Nefesh ❋ Ruach ❋ Neshamah

This is surely a Freudian source. As taught by traditional exponents of the Kabbalah, *Nefesh* represents your naked animal desires; *Ruach*, the middle soul that determines between good and evil; whilst *Neshamah* is your upper, immortal soul.

3 BOGATYRS

Alyosha Popovich ❋ Dobrynya Nikitich ❋ Ilya Muromets

The Bogatyrs were hero-knights who served at the court of the first Christian ruler of Russia, Vladimir I (958–1015), and fought under Alexander Nevsky. They can be compared to the Twelve Paladins of France and the Knights of the Round Table. *Alyosha* was the witty, dashing clever knight from Rostov; *Dobrynya*, the brave, courageous warrior from Ryazan;

and *Ilya*, the mystically inclined patriot from Muron. Part fictional, part historical, they have inspired numerous epic tales, films and patriotic visions – especially those articulated by nineteenth-century painters such as Victor Vasnetsov.

3 FOUNDING SIRES OF ENGLISH RACING

Darley Arabian ❖ Godolphin Arabian ❖ Byerley Turk

These three classic, thoroughbred horses were imported to England from the Near East between 1690 and 1728, after the Restoration had made gambling and horse-racing legal, and when Newmarket was emerging as the centre of this royal sport. The bloodlines of these Arab horses still dominate the racing world – indeed, 95 per cent of modern British thoroughbreds can be traced back to Darley Arabian

Byerley Turk painted by John Wootton.

alone. Eclipse, perhaps the greatest English racehorse of all time, was descended from both the Darley and Godolphin bloodlines and born during the solar eclipse of 1764.

3 CHILDREN OF LOKI

Hel ✳ Fenris ✳ Jormungand

Loki, the fire-like trickster god of the Norse, mated with the giantess Angrbo∂a to bring forth this alarming trio. *Hel* was a half-dead, permanently rotting giantess confined to rule the Underworld, Niffelheim, land of the unheroic dead. *Fenris* the wolf was destined to grow ever more monstrous before being magically chained by the gods who dwell in Asgard. (His sons Skoll and Hati would, however, free him from his chains by swallowing the sun and moon, before allowing him to attack and consume Odin at Ragnarok, the last battle.) *Jormungand*, the world serpent, would grow in power, hidden in the deeps of the ocean, until he embraced Midgard, the middle earth, by biting his tail. He was destined to do battle with Thor at Ragnarok, where Thor would crush him but after nine paces fall dead from a serpent's poisons.

3 ANCIENT DIETARY CLASSIFICATIONS

Ichthyophagoi (fish-eaters) ✳ Agriophagoi (game-eaters) ✳ Moschophagoi (eaters of roots and stalks)

These three ancient Greek classifications of the tribes along the Red Sea coast were based on food. And indeed food is still one of the essential identifiers of culture and identity. Laotians, for example, identify themselves as different from their Cambodian, Burmese and Vietnamese neighbours as people who love fish sauce, just as the English could once be called *rosbifs*; the Germans are still known as sausage eaters and the French attacked by American neocons during the Iraq War as 'cheese-eating surrender monkeys' (a term first used, in ironic mode, on *The Simpsons*).

3 GIFTS OFFERED BY A PHILETOR

Ox ❋ Drinking cup ❋ Suit of armour

In ancient Greece there was a tradition associated with the coming of manhood. An older man kidnapped a youth that he admired, usually with the assistance of others of his own age, and took him into the mountains to teach him how to be a man, to survive through his own wits, to hunt, sing, dance and fight – and, if so inclined, to love. At the end of this period, usually some two or three months, the young man would be returned to his city or village in honour, dressed as a citizen-soldier with the gift of an ox which he would sacrifice to Zeus, using his new cup to pour an oblation of blood, and later to drink wine, after the ox had been roasted and devoured in a communal feast with his own young friends. He remained the *parasatheis* (literally 'he who stood behind') to the older man, or *philetor* – a sort of godson-like role, his cupbearer at official feasts, as well as his companion on the battlefield.

3 JEWELS OF THE TAOIST

Compassion ❋ Frugality ❋ Humility

These jewel-like qualities are refined by some commentators as compassionate kindness, self-sufficiency and a form of humility that preserves life and refuses to direct others. They are defined in the Tao Te Ching, chapter 67: 'Here are my three treasures. Guard and keep them well. The first is pity; the second, frugality; the third, a refusal to see yourself as foremost of all things under Heaven. For only he that can pity is truly able to be brave; only he that is frugal is able to be truly generous and only he that refuses to be foremost of all things is truly able to become ruler of himself.'

3 GIFTS OF THE 3 MAGI

Melchior ❉ Caspar ❉ Balthazar

The much loved tradition of the Three Magi – Zoroastrian priests from the East – coming to visit the baby Jesus is only told in one of the Gospels, Matthew. No names or identities are given, but tradition records that *Melchior*, who came from Persia, brought a gift of *gold*, emblem of kingship and majesty. *Caspar*, who came from India, carried a gift of *frankincense* for the worship of the divinity. *Balthazar*, from Arabia, offered *myrrh* which was used in burial and anointing a sacrifice.

The Syrian Christians preserve a different set of names – Larvandad, Gushnasaph, Hormisdas – which suggests a much more accurate recall of the Magi from Zoroastrian Persia; as does the Ethiopian tradition with Hor, Karsudan and Basander. Each of the traditions, however, base their beliefs on Matthew's richly evocative description: 'They set out; and there, ahead of them, went the star that they had seen at its rising, until it stopped over the place where the child was. When they saw that the star had stopped, they were overwhelmed with joy. On entering the house, they saw the child with Mary his mother; and they knelt down and paid him homage. Then, opening their treasure chests, they offered him gifts of gold, frankincense, and myrrh. And having been warned in a dream not to return to Herod, they left for their own country by another path.'

3 FRANCISCAN ORDERS

Lesser Brothers ❊ Poor Clares ❊
Brothers and Sisters of Penance

Franciscan friars are customarily divided into three orders established by their founder Saint Francis: *Lesser Brothers*, Poor Ladies or *Poor Clares*, and a tertiary lay order – the *Brothers and Sisters of Penance*. Attitudes to the exact degree of poverty taught by Francis led to the Lesser Brothers being themselves divided between Friars Minor, Convectuals and Capuchins. Not, of course, to be muddled up with other orders such as the Carmelite, Augustinian or Dominican.

3 GODDESSES OF THE JUDGEMENT OF PARIS

Hera ❊ Athena ❊ Aphrodite

As the story goes, the trio were distracted upon being offered an apple labelled to the 'fairest of all'. They each sought to influence the judge – the prince-shepherd Paris of Troy – with the gifts of power, intelligence or the love of the world's

Botticelli's glorious Judgement of Paris.

most beautiful (mortal) woman. Paris chose the latter and set out to seduce Helen. She was, of course, married to Menelaus – an unfortunate detail that sparked the Trojan War.

3 PARTS OF AN ATOM

Proton (positive) ❋ Neutron (neutral) ❋ Electron (negative).

The *proton* is stuck like a plumb pudding together with its *neutron* partners, around which whiz the much smaller fly-like *electron* particles, within a space known as the electron cloud. This whole mysterious building block of life is held together by the power of electromagnetism to form atoms, which are listed in all their wonderful variety in that evocative list known as the Periodic Table of Elements.

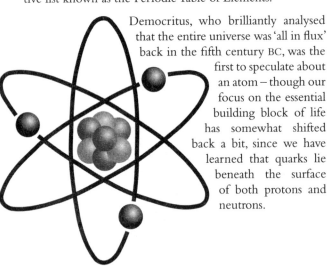

Democritus, who brilliantly analysed that the entire universe was 'all in flux' back in the fifth century BC, was the first to speculate about an atom – though our focus on the essential building block of life has somewhat shifted back a bit, since we have learned that quarks lie beneath the surface of both protons and neutrons.

3 PRECIOUS JEWELS OF BUDDHISM

Buddha (wisdom) ❋ Dharma (justice) ❋
Sangha (benevolence)

Wisdom is represented by the historical *Buddha* – the enlightened, the awakened one, as well as the sense of the highest spiritual potential of creation. The *Dharma* – 'that which is right' – are the ordained duties and teachings of Buddhism. *Sangha*, 'the assembly', can be the worldwide living community of Buddhists, those who have attained enlightenment, or a local community of monks and nuns.

3 PLAYWRIGHTS OF ATHENS' GOLDEN AGE

Aeschylus ❋ Sophocles ❋ Euripides

The apogee of Classical Athens' two-century-long golden age of literature was the generation who thought and wrote between 461 and 431 BC. Theatre-going Greeks of this time witnessed the noble high-minded and complex tragedies of *Aeschylus*, the graceful, measured characterisation of *Sophocles* and the more emotional and passionately charged creations of *Euripides*.

It is fitting that they are remembered as a trio, for each year three tragic playwrights produced a trilogy of tragedies (and a farcical comedy) that was performed over three consecutive days to honour Dionysus. These festivals were held around the time of the spring equinox. No more than three actors were permitted on the stage at any one time, their faces and that of the chorus covered in masks. At the end of the festival, one of the playwrights was voted the winner and given the prize of a goat, for the word 'tragedy' derives from 'goat song'.

3 PET HARES OF WILLIAM COWPER

Puss ❀ Tiney ❀ Bess

The company of three pet hares saved the poet William Cowper from madness in 1774. His favourite, Puss, died at 11 years and 11 months old, just after twelve noon. Hares are notoriously difficult to keep in captivity, due to their free spirit, so I have always thought that this stands as a great accolade to the way in which Cowper ran his house.

3 SONS OF ADAM

Cain ❀ Abel ❀ Seth

Cain and *Abel* and their fratricidal hatred are well-known. Not so *Seth*, the son 'granted' in place of the murdered Abel. Seth journeyed to Paradise for the olive branch of mercy for his father, Adam. In the gardens of Heaven he saw a child – destined to be born as the Redeemer of Mankind.

3 SONS OF NOAH

Shem ❀ Ham ❀ Japheth

The Book of Genesis records that Noah had his three sons in his 500th year – the Great Flood came in his 600th year. *Shem* was the legendary ancestor of the Semites of Arabia;

Ham, of the Hamitic people (whose homeland is North Africa and Ethiopia); and *Japhet*, of the black peoples of Central and West Africa. It was Ham, the youngest son, who derided his father for his naked drunkenness, leading Noah to curse Canaan – one of the oddest of all Old Testament stories, which was later used to justify slavery.

3 POWERS OF TIBET

Religion ✳ Government ✳ Tradition

A traditional Tibetan saying: 'The laws of our religion bind us with silk threads, the laws of the government are as heavy as a golden yoke, but the laws of national tradition are as inflexible as an iron pillar.'

3 REALMS OF THE NORSE

Asgard ✳ Midgard ✳ Niflheim

The Norse gods dwelt in *Asgard*, men in *Midgard* ('Middle Earth' – fashioned by Odin from the corpse of the slain ice-gaint Ymir) and the dead in *Niflheim*. All three realms were linked and upheld by the world tree, Yggdrasil.

THE ESSENTIAL TRINITY OF HINDUISM

Brahma ✳ Vishnu ✳ Shiva

Brahma, the Absolute, the Universal soul, can be worshipped in his three aspects of *Brahma* (Creator), *Vishnu* (Preserver) and *Shiva* (Destroyer). Brahma is often depicted with four heads sitting on a lotus; Vishnu is normally coloured blue and either lies on a many-headed cobra or rides Garuda, the man-eagle; Shiva is often depicted as the Lord of the Dance with necklaces of skulls, riding on Nandi, the bull.

The consort, or feminine form, of Brahma, is the goddess *Saraswati*, usually depicted in white on a lotus or sitting astride a peacock or swan; *Lakshmi* or *Bhoo Devi* is the maternal face of Vishnu; just as *Pravati* is the counterpart of Shiva, also herself known through a trinity of forms, as Shakti, Uma (the world mother) astride a tiger, or depicted as the terrifying goddess Kali.

THE TRISKELE

The Triskele – three running legs, often highly stylised – is a widespread ancient symbol. Like the swastika, it symbolises the continuous movement of the sun and/or moon across the sky, and is thus a badge of power, expressing the harmony

The Triskele as depicted on the Sicilian flag.

of the well-ordered heavens and trinity of divine powers. It appeared on some of the early coinage issues in Western Asia, on the shield of Achilles and as an emblem of Tyr the Norse sky god. In modern times, it remains the symbol and flag for three unlikely bedfellows – the Isle of Man, the island of Sicily and the Russian autonomous region of Ust-Orda Buryat Okrug.

3 WHEELS OF BUDDHISM

Theravada ❖ Mahayana ❖ Vajrayana

Sri Lanka, Burma and Thailand are the modern centres of the *Theravada* or southern Buddhist tradition, whose monks wear orange robes. Nepal, Bhutan, China and Japan follow the *Mahayana* or northern tradition, whose monks wear orange; its nuns wear black or grey robes. Central Asia, Mongolia and Tibet are the old spheres of the *Vajrayana*, whose monks wear red or maroon.

In addition, there are other schools of Buddhist thought known as Zen Buddhism, Shin or Pure Land Buddhism, as well as such modern offshoots as Nychiren.

3 DESTINATIONS

I'm going to Syria, said Reason; I'll come with you, said Rebellion ❖ I'm going to the desert, said Misery; I'll come with you, said Health ❖ Abundance said, I'm going to Egypt; I'll accompany you, said Humiliation

As quoted by the fourteenth-century Egyptian scholar Gamal Hamdan, showing how every advantage comes with its own innate disadvantage.

A gold coin of 313 AD showing Constantine with the Roman sun god Sol Invictus. Constantine decreed dies solis (Sunday) as a day of rest.

3 EMBLEMS WITHIN A CROWN

Power ❋ Legitimacy ❋ Victory

The Norman conqueror William I wore his crown three times each year: at Winchester at Easter, at Westminster at Whitsuntide and at midwinter at Gloucester. But, as Shakespeare tells us, 'uneasy lies the head that wears a crown'. For the crown stands for the three emblems of power, legitimacy and victory, but also for an ordained blood sacrifice as epitomised by the crown of thorns.

As an icon of power the crown has numerous lines of descent: the double crowns worn by the pharaohs of Egypt, the laurel wreaths of victory awarded to Greek heroes (and turned into the finest gold for Greek kings), the jewel-studded diadem worn on the brow by Persian and Hellenistic monarchs. The truest line of descent for the Western crown seems to have been the Greek radiant crown – Lucian's 'chaplet with sunbeams' – which was placed on statues of the sun god and which Constantine the Great co-opted in his fusing of the cult of the unconquered sun to the newly formed symbolism of a Christian Emperor.

3 ELIXIRS OF 'BROMPTON MIXTURE'

Cocaine ❖ Opium ❖ Brandy

Brompton Mixture was a popular remedy with which our grandparents gave relief from pain to the very ill and induced a cheerful sociability in the depressed or dying. It was named after the Royal Brompton Hospital in London, where it was dispensed to tuberculosis patients. The recipe had a number of variants, some aficionados adding chloroform water to the brandy, others championing heroin, while some argued that gin flavoured with cherry syrup more effectively masked the bitter taste of opium. All, however, agreed that the sum was better than any of the component parts.

2
=

OPPOSITES

Articulate the word 'Sun' and you soon find yourself thinking of the Moon. Man and Woman, Love and Hate, Right and Wrong, Farmer and Shepherd, Left and Right, Queen and King, North and South, Positive and Negative, Heaven and Hell, East and West, Life and Death, Victory and Defeat, Earth and Sky, Sunrise and Sunset. So, two is an inauspicious number in its cracking of unity.

ZOROASTRIAN DUALITY

Ahura Mazda ✹ Angra Mainyu

The ultimate duality is that of Zoroastrianism. The ancient Persian religion – dating probably to around 6000 BC – imagines a universe in creative conflict as *Ahura Mazda* (master of truth, light, spirit and goodness) struggles against *Angra Mainyu* (lord of the lie, darkness and evil). Thus the god of animating spirit is locked in eternal combat with the god of desire and earth-bound matter.

This concept was perhaps born from the even more ancient duality of the Indo-European deities, Sky and Earth (Asman and Zam), Sun and Moon (Mithra and Mah). The Prophet Zarathustra taught that the blood sacrifices and drinking rituals of the Aryans strengthened the dark powers and spirits of the earth (Daevas) and that one should instead aspire to worship the one true god with truth and righteousness, hymns and prayers. Every individual was given the free choice between the good path and evil, in the unending battle between truth and lies.

What makes Zoroastrian beliefs still so vivid is the sense of engagement as each worshipper consciously joins in this vital cosmic conflict of forces. Elements of this concept can be found in all other religions, and Zarathustra remains revered as a prophet by existing Zoroastrian communities, commonly known as Parsees, and by the Bahai and Ahmadiyyah communities. He was particularly honoured by the Prophet Mani, who preached his Manichean beliefs between the death of Christ and the birth of the Prophet Muhammad. In the West his dualist teachings influenced Saint Augustine (who was a Manichean as a young man), while the Cathars and Bogomils continued his belief system until their destruction as heretics.

LEFT BRAIN/RIGHT BRAIN, YIN AND YANG

Left Brain/Right Brain is the innate conflict within our own minds. It is also the creative balance between that part of our mind which rationalises, orders, creates processes and is logical, analytical and objective (the left brain) and that which is intuitive, thoughtful and subjective (the right brain). The creativity of an artist, a writer or an entrepreneur is a right brain concept, which requires a daring, free-spirited, imaginative, uninhibited, unpredictable and revolutionary mindset. The critical thinking required by an academic or an administrator needs the strengths of the left brain: reductive, logical, focused, conservative, practical and feasible. For anything to work well, there needs to be not only a balance but a fusion.

The most successful universal image of this is the T'ai Chi diagram: an egg composed of equal quantities of opposites: yolk and white, Yin and Yang. Yin is female, dark, earth-associated, passive, receptive and lunar. Yang is associated with male energies: light, Heaven, sun and the active principle in nature. Together they hatch mankind.

GUARDIAN PILLARS

Boaz ❊ Jachin

These are the two free-standing columns that stood before the holy of holies in the temple that King Solomon built for his god at Jerusalem. They were six feet wide and 27 feet tall, surmounted by gorgeous capitals decorated with hundreds of pomegranates draped with seven chains and topped with lilies. *Boaz* stood to the left, *Jachin* to the right.

....

THE 2 THINGS GAME

....

[1] People love to play the Two Things game, but rarely agree about what the Two Things are. [2] That goes double for anyone who works with computers.

A few years ago, Glen Whitman was chatting with a stranger in a Californian bar. When he confessed to this stranger that he taught economics, the drinker replied without so much as a pause for breath, 'So what are the Two Things about economics? You know, for every subject there are really only two things you really need to know. Everything else is the application of those two things, or just not important.' 'Okay' said the professor. 'One: Incentives matter. Two: There's no such thing as a free lunch.'

Inspired, Glen started playing the Two Things Game and recording some of the results on a web page (Google 'Whitman' and 'Two Things' and you'll get there). But it's more fun to try it for yourself – and especially good if you find yourself at a dinner next to a self-important professional,. Here are some of the best of Whitman's:

Finance: [1] Buy low. [2] Sell high.

Medicine: [1] Do no harm. [2] To do any good, you must risk doing harm.

Journalism: [1] There is no such thing as objectivity. [2] The end of the story is created by your deadline.

Theatre: [1] Remember your lines. [2] Don't run into the furniture or fall off the stage.

Physics: [1] Energy is conserved. [2] Photons (and everything else) behave like both waves and particles.

Religion: [1] Aspire to love an unknowable god. [2] Do this by trying to love your neighbour as much as yourself.

THE 2 COMMANDMENTS

Love the Lord ❁ Love Your Neighbour

The Gospel of Matthew records Jesus reducing the Ten Commandments to two golden rules:

Then one of them, a lawyer, asked Him a question, testing Him, and saying, 'Teacher, which is the greatest commandment in the law?' Jesus said to him: "'You shall love the Lord your God with all your heart, with all your soul, and with all your mind." This is the first and great commandment. And the second is like it: "You shall love your neighbour as yourself. On these two commandments hang all the Law and the Prophets."'

2 ANGELIC QUESTIONERS

Raqib ❁ Aatid

A pair of recording angels (Kiraman Katibin) sit invisibly on the shoulders of every believing Muslim, recording their doings, both good and evil. The recorder of good deeds, *Raqib*, sits on the right shoulder, while *Aatid*, who records bad deeds, sits on the left; mercifully, their books are only opened after your childhood is over. On the great day of judgement, it is believed that even your limbs will rise up and attest against you.

It is also said that a pair of terrifying angels – Munkar and Nakir ('The Denied' and 'The Denier') – visit the Muslim dead on their first night in the grave, who, after they have questioned them on matters of faith, move onto moral concerns. Just two questions are sufficient but your answers might require an eternity to fully confess. 'How did you acquire your wealth?' and 'On what did you spend it?'

OSIRIS'S TWIN SYMBOLS OF POWER

Flail ❋ Crook

The *Flail and Crook* of Osiris symbolise the two harvests achieved by the farmer and the shepherd and are one of the root sources for all symbols of power, notably the medieval orb and sceptre favoured by European royalty. In southern Asia, this is mirrored by the ritual sceptre (the Rodge) held in the right hand and a bell (the Drilbu) in the left of Indian statues. In Buddhist depictions the left hand may hold a Buddhist jewel, whilst the right hand is open in a gesture of sending blessings to the earth.

STAR-CROSSED LOVERS

Dido and Aeneas ❋ Helen and Paris ❋ Layla and Majnoun ❋ Antara and Abla ❋ Prince Khosrow and Shirin ❋ Pyramus and Thisbe ❋ Romeo and Juliet ❋ Abelard and Heloise ❋ Tristan and Isolde

Only the saddest love stories live for ever.

Aeneas would betray his lover *Dido*, the queen of Carthage (who had generously offered hospitality to his refugee-party from Troy) in order to follow his political destiny, while *Paris* would unwittingly start the whole gory cycle of the Trojan War by receiving the love of *Helen*, the most beautiful woman in the world, as reward from the goddess Aphrodite (see the *3 Goddesses of the Judgement of Paris*).

The love of *Majnoun* (literally the 'possessed' or 'mad one') for his beloved friend from school, *Layla*, is perhaps the most influential of all the Arab world's tales. The pair were separated by a family feud and after his beloved had been given to another man, Majnoun and wasted his life away in the desert, a virgin ascetic composing love songs to his impossible dream. Scholars have traced fifty-nine variations of this tale, including the cycle of *Antara and Abla*; the Persian story of the love of *Prince Khosrow* for Princess *Shirin*; *Pyramus and Thisbe*; and the most famous spin-off of all – *Romeo and Juliet* ('A pair of star-cross'd lovers take their life; Whose misadventured piteous overthrows, Do with their death bury their parents' strife').

Medieval European love was equally unpromising. The story of *Abelard and Heloise* begins with the elderly male canon-scholar seducing his brilliant but poor, young pupil in twelfth century Paris. Once pregnant she is sent away to give birth in Brittany and then tricked with a 'secret and private' marriage before being consigned to a nunnery. Only after Heloise's many admirers take their revenge on Abelard by castrating him does his proper love grow, and it is as chaste monk and nun that they enjoy the correspondence that would later be published.

Tristan and Isolde has inspired countless tellings, including Sir Thomas Malory's creation of the *Le Morte d'Arthur*. It has been traced to a twelfth century text but clearly looks back to a much older Celtic tradition in which the dashing young Tristan is sent to Ireland to bring back the beautiful Isolde for his uncle Mark, King of Cornwall. However, during their journey the two mistakenly drink a love potion destined to be consumed during the marriage ceremony. Thereafter their lives are full of deceit and romping adventure as they aspire to be good and dutiful to King Mark, yet stay true to their love. They can only break out of their fateful destiny by taking their own lives.

LOVER-MONARCHS

Antony and Cleopatra ✳ Justinian and Theodora ✳
Ferdinand and Isabella ✳ William and Mary

Antony and Cleopatra are the archetypal lover-monarchs. They first met at a magnificent conjunction of fleets off the coast of modern Turkey in the autumn of 41 BC. Antony was in command of the eastern half of the Roman Empire; Cleopatra ruled over the Hellenistic monarchy of Egypt; they met in order to forge a diplomatic alliance, but became lovers. Their attempt to conquer the East was destroyed by Octavian, but the pair gained immortality with their double suicides, their colourful descendants (Caligula, Nero and Queen Zenobia), and their leading Shakespearian roles.

The *Emperor Justinian's* long reign, which saw the definitive establishment of the Byzantine Empire, was aided by his trusted wife, *Theodora*, who brought a street-fighter determination to the partnership. Her mother had been a dancer and her father a bear-trainer, and she had grown up working in the circuses, brothels and dance halls of Constantinople.

Ferdinand of Aragon was a womanising, ruthless warrior-king of Aragon; *Isabella*, the intellectual heir of the richer but troubled Kingdom of Castile; they were cousins and their marriage began as an elopement. But their long reign was a political triumph, marked by their joint conquest of Moorish Granada (and notorious expulsion of Muslims and Jews) and the lucky patronage of Columbus and the discovery of America, which helped to forge the nation of Spain.

Britain's most significant joint monarchs were *William (of Orange) and Mary (Stuart)*: a personal union of cousins that ended the Anglo-Dutch naval wars and created a Protestant bulwark against Louis XIV's expansionist Catholic Kingdom of France. Their union allowed them to be 'jointly offered the throne' by Parliament when their uncle/father, James II, had been deposed. Mary miscarried their child in the first year of their marriage and was never able to conceive again, but kept an affectionate relationship with her husband, who had just one mistress and one boyfriend – his ex-pageboy Arnold van Keppel (who he elevated to Earl of Albemarle). The appeal of the Keppels as royal companions has remained constant, with Edward VII and, most recently, Prince Charles falling in love with Arnold's descendants.

PAIRS OF MALE LOVERS

Achilles and Patroclus ❈ Alexander the Great and Hephaestion ❈ Hadrian and Antinous ❈ Socrates and Alcibiades ❈ Zeus and Ganymede ❈ Apollo and Hyacinth ❈ James I and George Villiers ❈ Jonathan and David

Achilles and Patroculus are the devoted male lovers of Homer's Illiad – the grief of Achilles was so great that he would not bury the body until the ghost Patroclus begged him to do so in order that he might enter Hades. They were the role models for *Alexander the Great and Hephaestion* and for the Roman Emperor *Hadrian*, who, grief-stricken from his loss, declared his lover *Antinous* a god. Wise, but pug-faced *Socrates* adored the beautiful young aristocrat *Alcibiades*. Among the Greek gods, *Zeus* is eternally linked with his beautiful young cup-bearer *Ganymede*, whom he bore up to Heaven in the shape of an eagle, while *Apollo* had his *Hyacinth* (though the jealous love of the west wind, Zephyr, would conspire to make him his killer).

In more historic times, England's *King James I*, when chided for the public devotion he gave to *George Villiers*, the Duke of Buckingham, was able to answer, 'Did *Jonathan* not have his *David*? And did not David mourn him in words that we can read in the Bible?' Which indeed he did – the words

Achilles tends the wounds of Patroclus

being: 'Your love to me was more wonderful than the love of women. How have the mighty fallen, and the weapons of war perished.'

RIVAL BROTHERS

Cain and Abel ❋ Jacob and Esau ❋
Isaac and Ishmael ❋ Romulus and Remus

The story of *Cain and Abel*, the two sons of Adam and Eve, warns us of the jealousies that exist between brothers. Abel was the first shepherd, Cain the first tiller of the soil. But the murderous envy of Cain was inflamed when he saw that his brother's offering to God was deemed more acceptable as a sacrifice, so he killed Abel. *Jacob and Esau* did not murder each other, though Jacob tricked his firstborn brother of his birthright by selling him 'a mess of pottage' when he was hungry. The story of *Ishmael and Isaac* has its own tone, for the brothers were friends, but the elder would be driven from out of the tent of his father, Abraham, by his step-mother. *Remus* would be murdered by his brother *Romulus* during the foundation ceremony of the city of Rome.

NOAH'S PAIRS OF ANIMALS

As every infant knows, the animals went in two-by-two to the ark, as directed by Noah, on God's instructions ('bring into the ark two of all living creatures, male and female'). Those striving to apply a literal meaning to the Bible have, for many centuries, had a problem with ark logistics (dangerous animals, food and faecal matter) and of course, modern science has made the concept of pairs complicated. What about those that reproduce parthenogenically, or bees, where a drone and queen don't make a viable hive. And the pairs are complicated by Genesis itself, which also has God instruct Noah, 'take with you seven of every kind of clean animal, a male and its mate, and two of every kind of unclean animal, and also seven of every kind of bird'.

But the Great Flood is a great story – and, like the Garden of Eden, it is much indebted to the Sumerian *Epic of Gilgamesh*, in which the hero, Utnapishtim, builds a boat and sets off, more simply, with 'all the beasts and animals of the field.'

2-FACED JANUS

Two-faced Janus – the Roman deity of beginnings (hence the month January), doors, entrances and gates – looks both outwards and inwards, to the past and the future. The god, who had no Greek equivalent, presided over the beginning and end of conflicts, and the doors of his temple were left open in wartime and closed at peace.

1

YEAR 1

Our Western dating system – BC (Before Christ) and AD (Anno Domini – Year of Our Lord) – was conceived in the sixth century by a Romanian monk called Dionysius Exiguus, and came into widespread scholarly use after its adoption by the Anglo-Saxon historian the Venerable Bede. Prior to that, European historians dated years according to the Roman consul who held office in a given year.

Working in Rome, Dionysus declared that the current year was 525 AD, based on the birth of Christ taking place in the year 1 (there being no Western concept at the time of zero). Gospel historians later decided that Jesus was actually born a few years earlier, between 6 and 4 BC. Dionysius, it seems, may have wanted to disprove the idea that the end of the world would take place 500 years after the Birth of Jesus. That would have made it 6000 years after the Creation, which was believed to have taken place 5500 years before Christ. Dionsyius himself estimated, based on cosmological readings, that the end of the world would take place in 2000.

The CE/BCE (Common Era) designations, increasingly used to secularise history, are widely regarded as modern, politically correct innovations but were in fact introduced by Jewish historians in the mid-nineteenth century. But, for those who might want an alternative, there are plenty of other dating systems. The Jews start their calendar in 3761 BC; the Mayans, in 3114 BC; the Chinese, with the start of the Yellow Emperor's reign in 2696 BC; the Japanese, in 680 BC; the Muslims, with the emigration of the Prophet Muhammad from Medina from Mecca in 622 AD; the Copts, with the Year of the Martyrs in 284 AD, while the Ethiopian Church starts the clock back in 5493 BC.

NAMES OF THE ONE GOD

Abba ❋ Adonai ❋ Allah ❋ Atman ❋ Atun ❋ Brahma ❋ Christ ❋ Cosmic Intelligence ❋ Day of Brahma ❋ Dios ❋ Dieux ❋ Divine Light ❋ Elaha ❋ Ehyeh ❋ The Father ❋ The Full ❋ God ❋ Holy Wisdom ❋ Hu ❋ Huwa ❋ I am that I am ❋ Ishwar ❋ Jehovah ❋ Jesus ❋ Khoda ❋ King of Kings ❋ The Light ❋ Logos ❋ Lord of Heaven ❋ Lord of Hosts ❋ Messiah ❋ Om ❋ The One ❋ El Shadda ❋ Sovereign ❋ Tann Tao ❋ The Truth ❋ Universal Consciousness ❋ Varuna ❋ The Void ❋ Yahweh

The various names and ways of looking for God, the ways of addressing her and imagining him, saturate this little book. And if this eclectic gathering of numbers has got any one theme, or any point to it all, it is to show that mankind's spiritual vision has always been wonderfully rich and diverse but also unitary. One only has to read the all-inclusive name-lists of such pagan deities as Egyptian Ra or Babylonian Murduk to realise that all has always been seen in one. The different gods were no more than different-coloured windows looking out towards the one same source.

There is something almost comic about the energy of our various monotheistic faiths busily crushing all the dissenting altars and deviant schisms, but then quickly filling the gaps in our imagination with new hierarchies of revered teachers, saintly scholars and a buzzing spiritual stratosphere of guardian angels and divine messengers. Indeed, it is almost impossible not to recognise that the intolerance of a monotheistic priest is no more than a 'henotheism' – the worship of one specially favoured useful god before all others. For the true worshipper realises that even to try and express the concept of one is near impossible. For the one has to be both male and female, odd and even, past and future. The very moment we attempt to visualise 'one', we put ourselves outside it and create something other – we have formed the observer and the observed; we have made two.

0
══

ZERO

The mathematical sages of India conceived of this concept, and started on the path of the creation of zero back in the ninth century BC. The root of the word is the Sanskrit *Shoonya*, meaning 'it is a void', which got passed on to the Arabs, who knew the symbol as 'Safira'.

The Western world came to the concept extraordinarily late, when Venetian merchants stumbled across it and brought it back to their homeland, where it was known as 'zefiro' – later corrupted to 'zero'. The concept gradually spread through Europe, reaching such far-flung outposts as the British Isles in the late sixteenth century.

LOVE GAME

Scratch ❊ Duck ❊ Love ❊ Nil

Sport makes much use of the concept of zero, loading it with a multitude of names. There is *scratch* in golf, coined from 'scratching out' any trace on a score card. In cricket, a batsman who gets zero scores a *duck* – the slang for a bird that lays an egg, the shape of a zero. And that is the origin, too, of the word 'love' in tennis, corrupted from the English trying to copy the French for egg – *oeuf*. Football, meanwhile, favours the Latin *nil*, from *nihil* – nothing.

NIRVANA

The Sanskrit word '*Nirvana*' means 'blown out': a profound peace of mind, a freedom from suffering, and union with the Brahma-like symbol for the universe.

As the Lord Buddha explains, 'Where there is nothing; where naught is grasped, there is the Isle of No-Beyond. Nirvana do I call it – the utter extinction of aging and dying ... That dimension where there is neither earth, nor water, nor fire, nor wind; neither dimension of the infinitude of space, nor dimension of the infinitude of consciousness, nor dimension of nothingness, nor dimension of neither perception nor non-perception; neither this world, nor the next world, nor sun, nor moon. And there, I say, there is neither coming, nor going, nor stasis; neither passing away nor arising: without stance, without foundation, without support. This, just this, is the end of stress.'

ACKNOWLEDGEMENTS

A slim version of this book first appeared as a Christmas gift, sent out to friends of Eland Books. Then, some years later, I gave a copy to the publisher Mark Ellingham, over a cup of tea in our attic office, who thought it was the grit out of which a slim pearl could be created for Profile. The process of working with him to make this book over the last two years has been wholly enjoyable, with editorial sessions conducted at the hospitable tables of our locals, the Brill Café and Kick Bar in London's Exmouth Market. Many thanks to these two determinedly independent institutions for not moving their customers on and providing jugs of tap water a couple of hours after the last paid-for coffee.

I have recorded a comprehensive list of sources on my website (www.barnabyrogerson.com) but my personal slipstream of books that first and still inspired me in these lines of numerical enquiry comes from the works of James Fraser, Robert Graves and Joseph Campbell. I am also extraordinarily fortunate to have enjoyed the animating conversation of such mystery-questing beasts as Robert Irwin, Jason Elliot, Charlie Fletcher, Malcolm Munthe, Robin and Anne Baring, Rupert Sheldrake, Jill Purce, Mary Clow, Tim Mackintosh-Smith, Marius Kociejowski, Brigid Keenan, Alan Waddams, Mary Miers, Diana Rogerson, Rose Baring, Bruce Wannell, Kathy Rogerson, Ziauddin Sardar, Robert Chandler, Sylvie Franquet, Robert and Olivia Temple, Hugh Lillingston, Michael Scott, Paul Simpson, Duncan Clark, Anderson Bakewell and Ian Constantinides of the Ebenezer Chapel.

For the detailed work of hands-on editing I must repeat my thanks to Mark Ellingham and his team at Profile, especially designer Henry Iles, proofreader Nicky Twyman (who saved 1001 blushes), and publicist Drew Jerrison.

INDEX OF ENTRIES

Millions

Millions of Angels
Dancing on a Pin *1*
The 4,320-Million-
Year-Long Day *2*

Tens of 1000s

237,600 Miles or
30 Earths *3*
144,000 to be Saved *3*
124,000 Prophets *4*
84,000 Stupas of
Emperor Ashoka *4*
10,000 Blessings
of a Peach *5*
Xenophon's 10,000
Mercenaries *6*

1000s

6,585 Days of the
Saros Cycle *7*
Year One –
2696 BC *7*
1,460 Years of the
Sothic Cycle of
Ancient Egypt *8*
Ptolemy's 1,022
Stars *10*
1,003 Conquests of
Don Giovanni *10*
The 1,001 Nights *10*

Hundreds

666 – The Number
of The Beast *12*

540 Gates of
Valhalla *13*
The *Cinq-Cents* *14*
The 400 *14*
365 Days of Haab *15*
360 Degrees *15*
Vinaya – the 227
Rules *17*
114 Sura of the
Koran *18*
108 Stupas on
the Wall *18*
108 Names of
Krishna *19*
101 Names of
Ahura Mazda *21*
101 Dalmatians *21*

100

Power of 100 *22*
Let 100 Flowers
Bloom *22*
Homer's City of
100 Gates *23*

90s

99 Most Beautiful
Names *24*
99 Gold-Umbrella-
Bearing Rulers *26*
95 Theses of Martin
Luther *27*

80s

88 – Lucky in
China *28*

83 *Départements* of
Revolutionary
France *29*
82 Years of the
Buddha? *30*

70s

Two Sevens Clash *31*
77 Names of the
Great Hare *32*
74 Hidden Names
of Ra *33*
Seth's 73
Accomplices *34*
73 Benedictine
Rules *34*
72 Shia Martyrs *35*
72 Cathar Bishops *35*
72 Sacrifices to
Odin *36*
70 Names of
Jerusalem *38*
70 Cups of Poison *39*
70 Holy Idiots
of Sufism *39*
The Septuagint *40*
Gerald of Cremona's
Book of Seventy *40*

60s

66: The Number
of Allah *41*
The Kamasutra's 64
Arts of Love *42*
60 Degrees of
Sumeria *43*

50s

59 Regicides of
Charles I 44
56 Pillars 45
52 Shakti Pithas 46
The Mayan Calendar's
52-Year Cycle 46
Pentecost 47
Old Testament
Units of 50 47
50 Argonauts 48
Seven Sevens
are 49 50
49 Titles of the
Blessed Virgin
Mary 50
The Kabbalah's
42-Lettered God 52
42 – Life, the
Universe and
Everything 52

40s

42 Assessors 53
40 Days and 40
Nights 55
The Splitting
of a Hair Into
40 Parts 57
The *Arba'in* –
The 40 57

30s

The 39 Steps 58
33 – Number of
Completion 59
32 Signs of the
Universal Ruler 60
32 Grains of an
English Coin 61

The Greek League of
31 Against Persia 62
30 Days of
Ramadan 63
30 Pieces of Silver 63

20s

The *Canterbury Tales'*
29 Pilgrims 65
25 Prophets Within
The Koran 66
25 Wards of the City
of London 67
24 Tirthankaras
of the Jain 68
24 Nuba 69
The Age at which
the Prophet Mani
was Chosen 70
24 Angulas Make a
Forearm ... 71
... and 24 Palms
Make a Man 72
24 Letters of the
Greek Alphabet 73
The 23 Enigma 74
21-Gun Salute 75
20 Fingers and
Toes 76
Tolkien's 20 Rings
of Power 76

20–13

Sacred 19 of the
Bahai 78
Jewish Lucky 18 79
Sacred 18 of
the Whirling
Dervish 79
Hidden 17 of the
Bektashi 81

16 Prophetic
Dreams of Queen
Trishala 82
15 Ranks of the
Knights Templar 83
15 Men on the Dead
Man's Chest 85
Guaranteed Death
– Avoid 14 86
14 and Bach 86
14 Stations of
the Cross 87
13 at a Table – and
the 13th Month 89
13 Instruments of
the Passion 90
Satan's 13 Peers
of Hell 91
13 Hallows of
Britain 91
13 Bars on the
Union Flag 93

12

Power of 12 95
12 Signs of the
Zodiac 96
12 Months of the
Attic Calendar 98
12 Roman Months
of Our Year 98
The 12 Days of
Christmas 100
The 12 Days of
the Nowruz
Festival 101
12 Disciples at the
Last Supper 102
12 Sibyls 103
12 Olympian
Gods 105

The City's 12
Great Livery
Companies 106
Charlemagne's 12
Paladins 106
12 Knights of the
Round Table 108
12 Once and Future
Kings 109
12 Women of the
Prophet 110
12 Labours of
Hercules 112
12 Months of the
French Republican
Calendar 113
12 Symbolic Links
a Buddhist Must
Break to Escape
Suffering 114
12 Emblems
of Supreme
Authority 116
12 Sons of Pan 116
12 Feasts of Orthodox
Christianity 117
Leagues of 12 Cities
– Dodecapolis 120
12 Patriarchs 121
12 Tribes of Israel 121
12 Shiite Imams 122
12 Sons of
Ishmael 123
12 Precious
Stones 123
12 Chinese Character
Years 124

11

The 11th Hour 127
Dante's 11 127
11 Footballers 128
Cricket's First 11 130

10

2 Hands, 10
Fingers 131
10 Plagues of
Egypt 131
The 10
Commandments
132
10 Gurus of the
Sikhs 133
10 Magpies 133
10 Pains of Death 135
10 Deities Within
Bes 135
10 Sefirah of the
Kabbalah 136
10 Qualities
of a Bedouin
Warrior 137

9

A Pantheon of 9 138
9 Choirs of
Angels 138
9 Muses 139
9 Fire Altars of
Victory 140
9 Nights of Odin's
Sacrifice 141
9 Aztec Lords of
the Night 141
9 Mexican
Posadas 143
9 Rasa 143
9 Skills of a Norse
Lord 144
9 Freedoms of the
English
Declaration of
Rights 144
9 Worthies (Les
Neuf Preux) 145

9 Personalities of the
Enneagram 145
Chinese Power
of 9 146

8

8 Cherry Stones 148
8 Trigrams of the
I-Ching 149
8 Immortals 151
Eightfold Path of
the Buddha 152
Aristotle's Eightfold
Chain of
Politics 152
Eightfold Path of
Yoga 153
8 Lost Books of
Mani 153
Ogdoad 154
The 8 Greek
Winds 155
The 8 Gregorian
Church Modes 157

7

7 Ancient Visible
Planets 158
7 Notes 160
Fate of the 7
Days 160
Shakespeare's 7 Ages
of Man 161
7 Angels of
Punishment from
the Testament of
Solomon 163
7 Archangels 163
7 Chemicals of
the Alchemist's
Arcana 165

Ordering the 7
heavens *165*

7 Circuits of the
Kaaba *166*

7 Classical Heroes
Who Visited the
Underworld *167*

7 Churches of
Saint John *168*

7 Wonders of the
Ancient World *168*

7 Councils of the
Church *170*

7 Days of the
Week *171*

7 Deadly Sins of
Christendom *172*

7 Destructive Sins
of Islam *174*

7 Colours of
the Visible
Spectrum *174*

Snow White's 7
Dwarfs *174*

7 Hills of Rome *175*

The 7 Hausa
Cities *175*

7 Vowels *176*

7 Gates of Hell
Through Which
Inanna Surrenders
Her Divine
Regalia *176*

7 Senses *177*

7 Grades of Mithraic
Initiation *177*

7 Pleiades and
Hyades *179*

The 7 Liberal
Arts *180*

7 Ancient Things
for Nowruz *180*

7 Sages of Ancient
Greece *181*

7 Sacraments of
the Catholic
and Orthodox
Churches *182*

The 7 Seas *183*

The 7 Trials of
Travel *183*

Saptapadi: The 7
Steps of a Hindu
Wedding *184*

7 Valleys of the
Bahai *185*

7 Supreme Works of
Shakespeare *185*

6

Hexagons and
Ideals *186*

6-Pointed Star of
David *187*

6 Dynasties *188*

6 Confucian
Classics *188*

6 Zoroastrian
Immortals *189*

6 Days of Genesis *190*

6 Patrician Families
of Rome *192*

6 Evolutionary Stages
of History *193*

6 Mishnah of the
Talmud *194*

6 Physicians of
Antiquity *189*

6 Acts of Mercy *195*

6 Lines of the Last
Delphic Oracle *196*

6 Impossible things
of Gleipnir *197*

5

The Quincunx of
Heaven *197*

The Pentagram –
Solomon's Seal *197*

5 Precepts Taught
by Buddha *199*

5 Pillars of Islam *199*

The Khamsa (Hand
of Fatima) *200*

5 Books of the
Torah *201*

5 Colours of
Lungta *202*

5 Components of the
Soul in
Ancient Egypt *202*

5 Confucian
Blessings *203*

5 Favoured Meats
of the Desert *204*

5 Freedoms of
Psychoanalysis *204*

The Original 5-Man
Cabal *204*

5 Manifestations of
the Buddha *205*

5 Classical Orders *205*

5 Poisons of Old
China *206*

5 Precepts of
Apollo *206*

5 Qualifications of
Islamic Virtue *206*

5 Sufi Powers *207*

5 Rivers of Hades *208*

5 Regrets of the
Dying *208*

5 Restraints on
Sexual Passion *209*

5 Sacred Mountains
of China *209*

INDEX OF ENTRIES

5 Stars in the Flag of the People's Republic of China 210

5 Wizards in *Lord of the Rings* 210

5 Orders of Tree Nymphs 210

5 Reasons for Drinking 211

5 Virtues of a Turkic Beauty 211

5 Ways That Mutton Can Disappoint an Enthusiast 212

The Canaanite Pentapolis 212

The Cinque Ports 213

4

4 Elements 214

4 Cardinal Points 215

Ovid's 4 Ages of Civilisation 216

4 Arms of the Swastika 216

4 Suits of a Pack of Cards 218

4 Cardinal Virtues 219

4 Games of Ancient Greece 219

4 Pleasures of Omar Khayyam 220

4 Castes 221

4 Degrees of Attachment 221

4 Types of Caviar 221

4 Continents 222

4 Magical Possessions of Odin 222

4 Evangelists 223

4 Horsemen of the Apocalypse 224

Tetramorphs 225

4 Greater Prophets 226

4 Buddhist Pilgrimages 226

4 Symbols of Buddhist Enlightenment 227

4 Words of Buddhist Chant 227

4 Voices and String Quartets 228

The Quadrivium 228

4 Holy Marshalls of God 229

4 Celtic Seasonal Feasts 230

4 Doctors of the Church 230

4 Humours 231

4 Legions that Conquered Britain 231

4 Rightly Guided Caliphs 232

4 Rivers of Mount Kailash 232

4 Rivers of Paradise 233

4 Sons of Horus 233

4 Woods of the Cross 234

4 Sufi Questions 234

3

Divine Trinities 236

Tricolons 236

3 Fates 237

3 Furies 237

3 Graces 238

3 Gorgons 239

3 Goddesses of Pre-Islamic Arabia 239

The Tripartite Revolutionary Virtues 239

3-Leaved Shamrock 240

Clogs to Clogs in 3 Generations 240

3 Good Things of the Caliph's Dream 240

3 Virtues 241

3 Principles of a Good Zoroastrian 241

Freud's 3 Elements of Personality 242

3 Kabbalistic Components of a Soul 242

3 Bogatyrs 242

3 Founding Sires of English Racing 243

3 Children of Loki 244

3 Ancient Dietary Classifications 244

3 Gifts Offered by a Philetor 245

3 Jewels of the Taoist 245

3 Magi and their Gifts 246

3 Franciscan Orders 247

3 Goddesses of the Judgement of Paris 247

3 Parts of an Atom 248

3 Precious Jewels of Buddhism 249

3 Playwrights of
Athens' Golden
Age 249
3 Pet Hares
of William
Cowper 250
3 Sons of Adam 250
3 Sons of Noah 250
3 Powers of Tibet 251
3 Realms of the
Norse 251
The Essential Trinity
of Hinduism 251
The Triskele 252
3 Wheels of
Buddhism 253
3 Destinations 253
3 Emblems Within
a Crown 254
3 Elixirs of 'Brompton
Mixture' 255

2

Opposites 256
Zoroastrian
Duality 256
Left Brain/Right
Brain, Yin and
Yang 257
Guardian Pillars 258
The 2 Things
Game 259
2 Commandments
260
2 Angelic
Questioners 260
Osiris's Twin Symbols
of Power 261
Star-Crossed
Lovers 261
Lover-Monarchs 263
Pairs of Male
Lovers 264

Rival Brothers 265
Noah's Pairs of
Animals 266
Noah's Pairs of
Animals 266
2-Faced Janus 266

1

Year 1 267
Names of the
One God 268

0

Zero 270
Love Game 270
Nirvana 271

MORE ONLINE

I have posted online all the listings that had to be shortened for
this book (such as the 108 names of Krishna, the 101 names of
Ahura Mazda, the 99 ways to address the Virgin Mary, the 77
names of the Great Hare, the 74 hidden names of Ra, and the
full crew of 50 Argonauts), along with various curious facts that
didn't quite make the final edit, and a reading list of sources.

www.barnabyrogerson.com